THE

SECRET PERFUME

OF BIRDS

UNCOVERING *the* SCIENCE

of AVIAN SCENT

DANIELLE J. WHITTAKER

JOHNS HOPKINS UNIVERSITY PRESS

Baltimore

© 2022 JOHNS HOPKINS UNIVERSITY PRESS
All rights reserved. Published 2022
Printed in the United States of America on acid-free paper
2 4 6 8 9 7 5 3 1

Johns Hopkins University Press
2715 North Charles Street
Baltimore, Maryland 21218–4363
www.press.jhu.edu

Names: Whittaker, Danielle J. (Danielle June), author.
Title: The secret perfume of birds : uncovering the science of
avian scent / Danielle J. Whittaker.
Description: Baltimore : Johns Hopkins University Press, 2022. |
Includes bibliographical references and index.
Identifiers: LCCN 2021020601 | ISBN 9781421443478 (hardcover) |
ISBN 9781421443485 (ebook)
Subjects: LCSH: Birds—Odors. | Birds—Behavior.
Classification: LCC QL698.3 .W45 2022 | DDC 598.15—dc23
LC record available at https://lccn.loc.gov/2021020601

Cover and interior design by Amanda Weiss

A catalog record for this book is available from the British Library.

*Special discounts are available for bulk purchases of this book. For more
information, please contact Special Sales at specialsales@jh.edu.*

To Ellen Ketterson—
my mentor, role model, and friend

CONTENTS

....

CONTENTS

. . . .

FOLLOW YOUR NOSE

Follow your nose! It always knows!
—TOUCAN SAM, mascot for Kellogg's Froot Loops cereal

Follow your nose: To trust your own feelings rather than obeying rules or allowing yourself to be influenced by other people's opinions.
—Cambridge Dictionary

The bird was mocking me.

It was a beautiful Memorial Day weekend at an old-fashioned, lakeside mountain resort in Virginia. (In fact, it was the one made famous as Kellerman's in the movie *Dirty Dancing*.) I was perched precariously on the side of a hill that sloped down to the lake, where I had been trying for hours to catch this particularly elusive dark-eyed junco. Her mate, clearly less crafty than she was, had flown into my large but nearly invisible net shortly after I set it up a few feet away from their nest. I had measured him, taken samples, and safely released him over two hours ago. But this female, who went by the catchy name of 2221-574.23, seemed to have sniffed out the situation and had flown neatly around the net, easily evading capture every time she left or returned to her nest. I sighed and stared at a discarded Clamato can nestled among the pines.

I have never been particularly good at long-term planning. Mapping out an explicit five- or ten-year plan seems foolish to me because life never goes the way you think it will. I've found it better to have a general idea of what you'd like to achieve, and to then adapt to whatever unexpected events unfold along the way to your goal. Even if, like the male junco, you get trapped in a net and sampled for someone's study, your eventual release may encourage you to carry on in a different direction.

People who want to succeed in academia are typically advised to follow a well-traveled path, with a plan in place from the beginning. First, you should already know by the time you're in college that you'd like to be a scientist, and you should know what kind of science interests you. You should find an opportunity to conduct research in a lab while you are still an undergraduate, then know exactly what you'd like to pursue in graduate school and apply to work with a specific mentor. Once in graduate school, you need to carry out interesting, important, strategic research and publish your work before you graduate, which you must accomplish in six years or fewer. Then, you should have a postdoctoral fellowship already lined up, with clear plans for one or two years of research with another mentor. Finally, before this fellowship runs out, you should land a tenure-track position as an assistant professor and then spend the majority of your time writing winning grant proposals and training fabulous new graduate students and postdocs.

I did not do any of these things.

As an undergraduate at Emory University in Atlanta, I was a starry-eyed English major who understood nothing about how academia worked. I was valedictorian of my high school, with a perfect GPA, high standardized test scores, and a full scholarship to college. I wanted to be a professor of medieval English literature, and I assumed success in academia would follow the same formula as in high school: work hard, get good grades, then get the job you want. Also, I supposed that you would need to have some intriguing, original ideas—and I hoped that I would eventually develop some of those. But I was too shy and felt too far out of my element to ask for help. So, instead of actually meeting with professors and learning how academia worked, I simply followed the university's written guidelines about how to complete all the requirements for your major and for graduation. It seemed straightforward enough. But at some point, I realized that I should know quite a bit more about life in medieval Europe than I could glean just from studying the literature, and I decided to expand my education with medieval art history and Latin classes. I spent my junior year abroad at Trinity College in Dublin, Ireland—my first time ever leaving the United States—and there I intensively studied English literature. It was a dream made real. Only I discovered that it was not, in fact, my dream. Once I was exposed to the centuries of interpretation and research that scholars had already published, I started to feel that there was probably nothing fresh and compelling to be said about Chaucer or *Sir Gawain and the Green Knight*. And even if new insights were still to be had, I didn't think I was creative enough to have them.

When I returned home for my senior year, I felt confused and disenchanted with the route I had followed so far. I decided I needed to change everything, and I abruptly switched to a new major. Before my year abroad, I had fulfilled most of my science requirements with anthropology classes, and I had found them all interesting. Anthropology, as it is taught in the United States, is incredibly broad, encompassing cultural anthropology, biological anthropology, archaeology, and linguistics. In my senior year, I took fascinating courses to fill out the requirements for an anthropology major, including classes on nutritional anthropology, human sexuality, and—most important for the path I ultimately took—primate behavior and ecology. I loved learning about our closest living relatives, how they interact with each other, and how similar we are to them. I particularly loved gibbons, small monogamous apes of Southeast Asia with long arms who sing beautiful, mournful dawn songs.

When I graduated from Emory in 1996, I was no closer to figuring out what I wanted to do with my life than I had been at the beginning of college. I decided that academia probably wasn't for me. Every summer during college, I had worked as an office temp, impressing my employers with my organizational skills and typing speed. When a financial printing firm in Atlanta offered me a full-time position as a typesetter, I took it. The job paid well, especially for someone just out of college. Unfortunately, it was boring as hell, and soon I wanted to be anywhere else. I immersed myself in nonfiction books about primate fieldwork and fantasized about going to African or Asian jungles.

I honestly have no idea why I suddenly decided I should spend years in tropical rain forests, sweating and hiking and getting bitten by mosquitoes. I was never outdoorsy as a child. I went camping a couple of times with friends, and sure, I wandered through the woods a bit on my own. But looking back at those times, I see no evidence of a budding young naturalist. I was, instead, indoorsy: I liked to read books and listen to music and watch television. I was not active—in the eighth grade, I failed gym class because I refused to participate. But I was dissatisfied at the prospect of living the mundane life that all my peers were embracing. I had no interest in desk jobs, real estate, or raising children. I wanted to travel and experience things that ordinary people never did—and these stories about fieldwork showed me a different way to live.

I surreptitiously began applying to PhD programs in biological anthropology. Why "surreptitiously"? I wasn't ready to tell my parents. Though neither of my parents had a college degree, they had always pushed me to do well in school and pursue higher education. For the most part, this was a good thing, but sometimes they went too far. If I brought home a test with a grade of ninety-seven, my father would say, "What happened to the other three points?" At our celebratory dinner the day I graduated with my bachelor's degree from Emory, my dad—who never finished high school—cheerfully exclaimed, "Now you just have to get your PhD!"

In that moment, it seemed that nothing I did would ever be quite good enough. Inside, I was devastated, and my immediate response was defensiveness. Throughout the following summer, I denied my growing unease, insisting that my office job was fine, that I didn't need to get another degree, and that I just wanted to live a conventional life. And I tried to believe that it was true.

When I finally admitted to myself two years later that I wanted to go to graduate school, it felt oddly like failure. I told very few people and did not seek out advice. I applied to several graduate programs, but coming from my rather haphazard educational background, I didn't really understand the system. The PhD program I was most interested in was the New York Consortium for Evolutionary Primatology (NYCEP), which combined training from the Anthropology Departments of Columbia University, New York University, and the City University of New York. I applied to all three universities, not caring which one I got into. Typically, if faculty members are interested in mentoring you as a graduate student, they invite you to visit and learn more about the program. But I didn't know that. Instead, I arranged to take a trip with one of my closest friends to the city, and I contacted several professors in the program to let them know I would like to meet with them. Ultimately, my clueless persistence paid off: I was accepted into the graduate program at the City University of New York, as part of the NYCEP program. I even got a small fellowship, thanks to my good grades in college and high GRE scores.

As soon as my first semester of graduate school began, I worried that I had made a terrible mistake. The other students all seemed to have a stronger academic background, a better understanding of how things worked in graduate school, and clearer ideas about their research plans than I did. I vividly remember spending one evening in the lab with an array of primate and mammal skulls spread out before me, trying to figure out the difference between molars and premolars. Teeth play a very important role in identifying fossils, and I needed to learn how to identify them in my Evolutionary Morphology class. Molars, which are furthest back in the mouth (and are used for grinding and chewing), are larger and have more cusps than premolars. But alone in the lab, the worn, off-white teeth all looked the same to me. I burst into tears. I was sure that I would never be able to see the things that must be obvious to everyone else in the program. I was sure that I didn't belong there.

Eventually, I did learn to distinguish between a molar and a premolar—and I also learned that I was most definitely not alone. None of the other graduate students felt like they knew what they were doing either. No one believed they belonged, and they all thought everyone else knew more. Even the arrogant know-it-all "mansplainer," who enjoyed correcting professors when they misspoke (all programs have at least one), turned out to secretly fear he was inadequate.

I survived graduate school, although it took me seven and a half years to complete my degree. I was able to realize my dream of following gibbons in the tropical forests of Indonesia. I developed a research project on the evolutionary genetics and conservation biology of the Kloss's gibbon, a small, endangered, and poorly understood ape found only in the Mentawai Islands of Indonesia. I spent a total of a year in the Indonesian rain forest, and several more years in the laboratory coaxing DNA out of gibbon poop. To pay the bills, I taught many undergraduate anthropology and biology classes as an adjunct lecturer, though the tiny paychecks barely covered the rent of our one-bedroom apartment in the Bronx. I spent another year in New York after I graduated, patching

A young me on North Pagai Island in the Mentawai Islands, Indonesia, 1999.

together part-time teaching gigs at multiple colleges while applying for full-time faculty and postdoctoral positions.

Finally, I found the job that changed my entire career trajectory: a position as a postdoc studying bird behavior at Indiana University. Which is how I ended up on the side of a hill in Virginia, trying to capture a recalcitrant junco. I did eventually catch her. I collected the data I needed, released her, packed up my equipment, and headed back up the mountain to the research station for Memorial Day beers with my labmates. It had been a tough day.

To be honest, I still don't really know what I'm doing. I'm still afraid that someone will figure out how clueless I am and make me go back to a boring regular life. I feel incredibly lucky to have gotten where I am today. Following my meandering path of scientific inquiry has been extremely rewarding and has kept me going for decades now.

In this book, I want to share not only the fascinating science of how birds use their sense of smell—a topic I certainly never expected to become the focus of my life—but also the reality of

The wiley junco that took me more
than two hours to catch, a.k.a.
2221-57423

what it's like to be a scientist.* The story of how birds smell—my
story—is proof that even though plans often don't play out as expected, the results can reveal new and more stimulating paths. I
hope I can convince you how important it is to question long-held
assumptions, in science and in life. And most of all, I want to in-
spire potential future scientists who might otherwise be intimidated by the seemingly staid, precisely plotted world of science.
The truth is, many of us are just following our noses.

---

* I also plan to share my love of discursive footnotes, and I hope you won't mind too
much.
Note: When you see italicized terms in the text, it means they are explained more
fully in the glossary at the end of the book.

THE SECRET PERFUME OF BIRDS

CHAPTER 1

THE MOST ANCIENT AND FUNDAMENTAL SENSE

It has become a truism in the past century that the subject of avian olfaction is a controversial one. Each of the occasional research articles, notes, and reviews that has appeared, ranging through the fields of ornithology, physiology, anatomy, and psychology, typically has begun with the statement that the olfactory ability of birds is open to dispute.
—BERNICE WENZEL, 1967,
"Olfactory Perception in Birds"

BIRDS CAN'T SMELL

"Birds don't have a sense of smell, so I don't understand why you'd study that anyway."

This extraordinary statement, expressed offhandedly by neurobiologist Dr. Jim Goodson while we waited in a cafeteria line at lunchtime, caught me off guard. Every form of life, even plants and bacteria, has the ability to sense chemical compounds in their environments. Chemical senses, which include smell and taste, are critical for avoiding harmful substances, like poisons, and find-ing beneficial ones, like food. Yet here was a well-respected biolo-gist telling me that an entire class of animals, encompassing nearly twenty thousand species, lacked what is often called "the most an-cient and fundamental sense." That couldn't be right, could it?

I was a postdoctoral researcher in the Biology Department at Indiana University, and that afternoon, I was casually chatting with Goodson about the difficulties I was having in the lab. I was study-ing dark-eyed juncos, gray and white sparrows that are common throughout North America. I was interested in why they choose a specific individual to mate with, and why sometimes they are loyal to their mates but other times they cheat. I was specifically at-tempting to explore the role of a family of immune-related *genes* called the *major histocompatibility complex*, or MHC for short. MHC genes had been the subject of much debate in the previous decade or so. Although the primary role of the products of these genes is to detect potentially harmful invaders, such as bacte-ria and parasites, researchers suspected that MHC might be the basis for sexual attraction in many animals, and maybe even in humans. Excited by the possibility of resolving the mysteries of mate choice, I dove headfirst into the project. In 2008, MHC genes had not been studied much in birds, but in general, animals were thought to detect MHC by *olfaction*, or the art of smelling. The du-bious Dr. Goodson was intimating that since birds couldn't detect scent, MHC was probably inconsequential in their mate choice decisions, and therefore wasn't worth studying.

Although I had a decent education in evolutionary biology, my PhD research was in primatology, and I was still new to ornithol-ogy. I was constantly surprised by all the ways that birds were dif-ferent from mammals. For example, most female mammals have two functioning ovaries, one on the left side and one on the right side of the body. But in birds, only the left ovary develops, which reduces overall body weight. Compared to mammals, birds have more efficient circulatory and respiratory systems, allowing them to direct more of their energy to flying. In fact, most of the differences I knew about were adaptations to flight. These changes made intuitive sense to me, and they were clear examples of how evolution works: a trait that increases an animal's ability to survive and reproduce becomes more common because those survivors pass it on to more descendants. Similarly, traits that decrease an animal's success are less likely to be passed on because animals with those traits don't survive as long or have as many offspring. Yet losing an entire sense didn't seem to me like it would improve the odds of anyone's survival! Surely, not being able to smell would be a big disadvantage, since smell is important for sensing the environment around you.

Because my ornithology textbooks were quiet on the topic of bird olfaction, I started scouring the literature looking for evidence to support Goodson's seemingly counterintuitive claim. Soon, I found that beliefs about *anosmic* birds—that is, birds without a sense of smell (from the Greek *osmē*, "odor")—had been around for decades, although they were rarely mentioned in scientific literature. Neurobiologists like Goodson noted that the *olfactory bulb*—the part of the brain that receives information from receptors in the nose—is unusually small in birds. However, not all birds display this trait: for example, it is widely accepted that turkey vultures are attracted to the scent of carrion. Also, the "tube-nosed" seabirds, so called because of the shape of their nostrils, have relatively large olfactory bulbs and can find food at sea using scent. Kiwi birds in New Zealand are nocturnal but have poor eyesight, so they use scent to detect insects and worms in the dark. While these anomalies were acknowledged in the books, they were presented as exceptions to the rule that birds had little to no sense of smell.

The conventional wisdom stated that birds gave up the ability to smell in exchange for superior eyesight. Indeed, most birds have exceptional eyesight, better than any mammal. Raptors have excellent visual acuity and can see over very long distances— eagles can detect the movements of small prey animals from a great height, and owls have evolved especially large eyes so they can see even in very dark conditions.

In addition to aiding in hunting, eyesight is important in mate choice. Some of the flashiest animals in the world are male birds— think of peacocks and birds of paradise. These birds sport ornate tails and crests showing off brightly colored feathers, primarily for the purpose of attracting females. While extravagant plumage is a commonly recognized feature of birds, less well-known is that birds can actually see more colors than mammals. Most birds have tetrachromatic vision, meaning that they can see four colors, using four different types of cone receptors in their retinas. Humans and most other primates are only trichromatic, with red, green, and blue receptors. Birds' fourth type of cone receptor allows them to see colors in the ultraviolet wavelengths, which means that some feather colors that look dull to us are actually much more intriguing and attractive to a bird. Ultraviolet sensitivity also allows birds to see more clearly and navigate better in dense foliage, as individual leaves stand out more because they reflect ultraviolet light. These improved visual abilities have obvious advantages: predators with better eyesight will be more successful in obtaining food, and males with more flamboyant plumage will mate with more females and sire more offspring.

But why would evolving improved eyesight come at the cost of smell? What disadvantage would necessitate such a trade-off rather than simply enhancing one sense? The concept made no sense to me, and I couldn't find any scientific explanations, only assertions. This widespread acceptance of an unsupported "fact" rankled me. Suddenly, I knew I had a new adventure to pursue, one that changed the course of my research and my life.

ANTI-NOSARIANISM

Birds' supposed lack of smell was cited everywhere from undergraduate animal behavior textbooks, to a falconry workshop my husband attended, to internet memes assuring the public that it's ok to return a fallen baby bird to its nest because its mother "cannot smell." But no scientific study had ever demonstrated that birds don't have a sense of smell. To understand where this belief originated, it's helpful to start with the basics.

How does the sense of smell work? The first step in smelling an odor is to inhale air containing small organic molecules. Receptors in the lining of the nose bind to these molecules, activating olfactory sensory neurons that carry impulses to the olfactory bulb at the front of the brain. The olfactory bulb looks like two long ovals, one situated just above each nostril. Certain parts of the brain, such as those associated with memory, connect to the olfactory bulb, allowing us to recognize the odor. A common way to compare the olfactory capabilities of different species is to examine the sizes of their olfactory bulbs. You can compare either the absolute size of the olfactory bulbs, or—better—the size of the bulb relative to rest of the brain (in other words, what percent of the brain is made up by the olfactory bulb). Animals with larger olfactory bulbs are thought to have a more highly developed sense of smell and to rely more heavily on scent than those with smaller bulbs. In birds, the olfactory bulb is extremely small relative to the size of the rest of the brain, at least when compared to mammals. One study that cataloged the brain anatomy of dozens of bird species was frequently cited as proof that birds had limited smelling abilities, yet the intent of the study was to allow comparison among bird species, not comparison with mammals.

Studies focusing on the size of the olfactory bulb provide no information about how well it works. I knew from my introductory ornithology reading that avian brains have significantly less white matter (tissue that enables brain cells to send and receive information) than mammalian brains—and yet birds are able to use their brains just as efficiently. Here's the twist: although bird brains are smaller than mammalian brains, the neurons are much more tightly packed. Some birds are capable of complex cognitive functions despite lacking a neocortex. The neocortex is a structure in the forebrain thought to be the source of higher cognitive functions, and it is found only in mammals—the largest are found in humans. Yet some birds, like crows and other corvids, display evidence of sophisticated cognitive functions, including insight, the ability to suddenly gain knowledge of a solution to a problem. Much of the forebrain in birds is structured very differently from that of mammals, but birds can perform many of the same cognitive functions. So why wouldn't we expect the avian olfactory bulb to also have evolved more efficient processing?

The credit for the myth—or perhaps the blame—goes mostly to John James Audubon, the great cataloger and illustrator of birds. In the early nineteenth century, as today, turkey vultures were believed to use their sense of smell to guide their scavenging efforts. After noticing that vultures frequently failed to detect hidden food, Audubon set out to test this assumption. His opinions on the topic are immediately clear from the title of his 1826 paper: "Account of the Habits of the Turkey Buzzard (*Vultur aura*) Particularly with the View of Exploding the Opinion Generally Entertained of Its Extraordinary Power of Smelling." Audubon conducted three experiments to test whether the vultures were using sight or smell to find food. First, he stuffed a dead deer with grass and let the flesh dry completely, so that it gave off no smell of rotting flesh. He then left it out in the middle of a field where animals could easily see it. The vultures attacked the deer carcass, apparently taking quite a while to realize that there was actually no meat left inside. Second, he concealed a dead pig carcass in the heat of summer and waited to see whether the vultures located the rotting meat. (Technically, he did not do the concealing himself. In his own words, "I made the negroes conceal the hog." Although it's rarely acknowledged, Audubon held enslaved people.) Many vultures flew over the area in search of food, but while dogs found and ate the pig carcass, the vultures never approached it. Finally, he hand-raised two vulture chicks and approached their cage with meat in his pockets. The chicks showed no reaction to the smell of the meat alone, and they only reacted when they could see him. From these experiments, Audubon concluded that turkey vultures used sight, not smell, to look for food.

This paper sparked decades of controversy between two factions that one chronicler dubbed "Nosarians" and "Anti-Nosarians."* A prominent Nosarian, Charles Waterton, in his 1837 book, *Essays on Natural History, Chiefly Ornithology*, wrote a highly critical analysis of Audubon's experiments. Waterton's critique of every aspect of Audubon's paper was so scathing that one book reviewer, in the *London Quarterly Review*, felt compelled to comment that he should be kinder: "[Waterton] should remember how prone we all are to error, and that we should be a little tolerant of those who do not happen to think exactly as we do." Of the first experiment, in which Audubon claims he fooled the vultures with a stuffed deer carcass, Waterton wrote: "I have a better opinion of the vulture's sagacity, than to suppose that he would have spent so much of his precious time upon the rudely stuffed mockery of an animal, unless his nose had given him information that some nutriment existed in that which his keen and piercing eye would soon have told him was an absolute cheat." About the second experiment, in

---

* These names were bestowed in an article in the *London Quarterly Review* by an anonymous reviewer of Charles Waterton's *Essays on Natural History, Chiefly Ornithology*. The names refer to a story in Laurence Sterne's eighteenth-century novel *Tristram Shandy*, in which religious sects, called the Nosarians and the Anti-Nosarians, debated whether God could create an infinitely large nose.

which Audubon hid a rotting hog, Waterton pointed out multiple problems: "Here the author positively and distinctively tells us, that he saw *many* vultures, in search of food, sail over the ravine, *in all directions*, but none discovered the carcass; although, *during this time*, several dogs had visited it, and fed plentifully on it. Pray, when the dogs were at dinner on the carcass, and the vultures at the same time were flying over the ravine where the hog lay, what prevented these keen-eyed birds from seeing the hog?"

People on both sides of this scientific dispute made many creative (and sometimes cruel) attempts to demonstrate that birds either could or could not smell. In one such experiment, the researcher put ammonia in water to see whether pigeons would avoid drinking it. Rather than staying away from the tainted water, the pigeons tasted it and "jump[ed] away horryfied." In another, a researcher attempted to train ring doves to follow a scent to find food in a labyrinth (they failed). Still another tried to train songbirds to respond to odors by giving them electric shocks when they were wrong (sometimes the training was successful, sometimes not). And, of course, many repeated the ever-popular trick of hiding food in various stages of decomposition to test whether vultures uncovered and ate it.

Ultimately, it seems that the biggest problem with Audubon's experiments is that he wasn't even studying turkey vultures. His descriptions of the birds and their behavior indicate that he was actually working with black vultures, a different species. Although turkey vultures do indeed rely on scent to find food, black vultures have keen eyesight to detect prey. Even so, turkey vultures are known to avoid overly rotted meat, preferring carcasses that have only been dead for up to four days. But beyond these significant errors, the underlying problem with all of the studies is that researchers were asking the wrong questions, muddling the answers, and coming to unsupported conclusions.

THE NEW LONDON SCHOOL EXPLOSION

In 1937, the southeast Texas town of New London boasted one of the richest school districts in the country. Despite the economic troubles of the Great Depression, New London had built a beautiful, modern school in 1932. Among other features, the school had the first lighted football stadium in America. New London was able to afford these luxuries thanks to the discovery of a particularly productive oil field in their county seven years before.

Setting aside the boiler and steam heat systems that were in the original plan, the builders installed seventy-two natural gas heaters throughout the school. However, they did not adapt the basement area to provide adequate ventilation. The school's natural gas was provided by the Union Oil Company until January 1937, when the school canceled the contract and instead tapped into the free residue gas line from Parade Gasoline Company. Natural gas that is extracted along with oil during drilling is considered a by-product and is flared off. At the time, it was very common for individuals, schools, and churches to take advantage of this waste product, consuming it as a free source of energy. This practice was unregulated, and gas companies generally turned a blind eye since there didn't seem to be any harm to it.

Natural gas is odorless and colorless. So, when the gas line began leaking and filling up the 253-foot-long, 56-foot-wide crawlspace that ran underneath the entire length of the school, no one detected it. By mid-March, students had been complaining about headaches for several days, but no one had paid much attention.

On Thursday, March 18, 1937, the school was preparing for a major academic and athletic competition that was to take place the next day in a nearby town. The students in first grade through fourth grade had been let out early, and school buses were in the process of taking them home. The fifth grade through eleventh grade students, at least five hundred of them, were still in classrooms at three o'clock, awaiting the final bell that would dismiss them. That bell never rang.

At 3:05 p.m., a shop teacher turned on a sander in his basement classroom. A spark from that machine ignited the gas that filled the space underneath the school. The result was a massive explosion. According to witnesses, the walls visibly bulged, and the roof lifted off of the building before it crashed back down, collapsing the building and killing more than three hundred people, mostly children.

This explosion remains the deadliest school disaster in US history. Just a few weeks after the explosion, the Texas State Legislature enacted a law requiring natural gas companies to add mercaptans—smelly, sulfur-containing organic compounds—to natural gas so that leaks could be detected by smell.

Two decades later, ornithologist Kenneth Stager at the University of Southern California was working on resolving that long-standing debate over the turkey vulture's sense of smell once and for all. While trying to determine the best method to test whether these birds were attracted to the scent of carrion, he met Ralph Openshaw, an engineer from the Union Oil Company, and learned that the field engineers had long recognized the usefulness of turkey vulture noses. Two years after the New London school explosion, the Union Oil Company was struggling to locate leaks in a section of a natural gas line between Orcutt Hill and Avila, California. According to Openshaw, an engineer from Texas suggested that they add a large amount of ethyl mercaptan into the line and wait for the turkey vultures to circle around the area of the leak. As it turns out, ethyl mercaptan is a key component of the scent of carrion: the favorite food of turkey vultures.

Stager first conducted pilot tests in an area where he had observed turkey vultures roosting. He simply opened a container of ethyl mercaptan upwind of their flight path and waited with a spotting scope. A mere twenty minutes later, three vultures were tightly circling the area above the container. Two more wildly successful field trials convinced Stager that ethyl mercaptan was an appropriate attractant for testing the birds' sense of smell, and he developed a more sophisticated compressed air unit to dispense controlled concentrations of the compound. In his doctoral thesis, Stager meticulously describes multiple trials that began at nine o'clock in the morning and kept vultures circling his device until the gas ran out in the early evening.

It appeared that Audubon was wrong. Not only were turkey vultures attracted to a scent found in carrion but they were also able to find the source of the scent without any visual cues.

HOW TO ASK THE RIGHT QUESTIONS

How can you determine whether an animal smells an odor? In the mid- to late twentieth century, UCLA neuroscientist Dr. Bernice Wenzel suggested that to really understand if and how birds use their sense of smell, we need to break down the questions into three parts. First, do birds perceive odors? If we determine that they do, then next we ask whether they change their behavior in response to odors. Finally, if they do not change their behavior in response to an odor, can they *learn* to change their behavior? Wenzel argued that the experiments by the Nosarians and Anti-Nosarians of the previous century had been designed to test only questions two and three (do/can they change behavior in response to a smell?), but then used the answers to conclude that the answer to question one (can they smell?) was negative.

Wenzel was a pioneer in the field of avian olfaction, researching bird brains, olfactory structures, and neurophysiology. She pursued these three questions across a range of bird species. If an odor has any sort of biological significance to that animal—if it indicates a food, a mate, or a predator, for example—then the body should produce an appropriate reaction. Wenzel presented birds with odors and measured physiological responses, like increases in heart rate. Wenzel's research led to the recognition of many of the birds whose olfactory abilities are now considered exceptions to the alleged anosmic rule.

The brown kiwi, a nocturnal bird native to New Zealand, forages for worms and insects in the dirt with its long, thin beak. Because kiwis have tiny eyes and weak eyesight, it was thought that they might use their sense of smell to find food. Wenzel set up feeding stations in aviaries where the captive kiwis ate food from tapered aluminum tubes that were sunk into the ground, mimicking their natural feeding habits but providing the researcher with feeding dishes that could be manipulated. The birds' regular diet consisted of a "redolent" mixture of raw steak, raisins, and starter mash. In a particularly convincing experiment, Wenzel put this food mixture into one of the three feeding tubes and nothing but dirt in the other two. She then covered the tubes with a fine-mesh screen that the birds could pierce with their beaks, but she left some space between the screen and the contents, so that the only way the kiwis could tell the difference between the food tubes and the dirt tubes was by scent. In every test, the kiwis opened all of the tubes containing food but none of the dirt-only tubes.

In another set of experiments, Wenzel added cotton saturated with an artificial scent like amyl acetate (fake banana) or synthetic musk to the bottom of the food tubes. The foreign scents made no difference to which food tubes the birds preferred. The lack of any response to these artificial odors, despite the other experiments demonstrating the birds' olfactory capabilities, shows the importance of finding a relevant way to ask your research question. Researchers using synthetic scents to test birds' ability to detect odors may get negative results simply because these odors are not relevant to the birds.

Wenzel also tested the birds' physiological responses to a gentle air stream scented with various odorants (including food scents and synthetic compounds). Her measurements included electro-encephalogram (EEG) readings, which measured electrical activity in the brain, and respiratory rate, to measure how fast the bird was breathing. When presented with odors, the birds' respiratory rate increased, and their EEG readings changed from a typical drowsy, inattentive pattern to a state of alertness. These physiological responses suggested that the birds did indeed detect the odors. The EEG patterns showed that the scents were being processed in the brain, and the changes in other vitals suggested a real reaction to scent stimuli, and not just to air blown in their faces.

With collaborator Larry Hutchison, Wenzel showed that tube-nosed seabirds could locate food using only scent. Wenzel and Hutchison sailed off the coast of California, bringing along containers of liquids that emanated either food-related odors (like tuna oil, bacon fat, and ground-up squid) or non-food-related odors (including motor oil, mineral oil, and ocean water). These liquids were put out either as an open slick on the water or on a floating raft. They found that, overwhelmingly, these seabirds were attracted to the food-related scents and not at all to the non-food-related odors, and that the birds showed distinctive foraging patterns, including approaching the scent from downwind and following it in a zigzag pattern. Wenzel and Hutchison attracted sooty shearwaters, pink-footed shearwaters, northern fulmars, and black-footed albatrosses, plus several others that appeared less often. These birds showed up across different seasons and in all kinds of weather conditions, and they attempted to land directly on or near the odor sources.

Meanwhile on the other side of the country, Betsy Bang, a medical illustrator at Johns Hopkins University, became interested in avian olfaction while drawing dissections of nasal cavities for her husband's scholarly articles on respiratory disease in birds. Observing how large and complex these olfactory structures were in some birds, she suspected that the widespread opinion of birds' poor sense of smell was incorrect. Bang dissected the nasal cavities of three different kinds of birds: the turkey vulture; the nocturnal Trinidad oilbird, which nests and forages in completely dark caves; and two species of albatross, which forage over open water. These birds are completely unrelated and have very different nesting and feeding habits, yet Bang showed that they all have noses that seem built for smelling. She showed that the nasal passages of these birds have a broad surface area enabling the birds to "filter, warm, moisten, and sample inspired air chemically." Within the passages, she found scrolled conchae—hollow spaces that provide additional surface area for the nasal mucosa. In some birds, like the turkey vulture, these structures are so elaborate that they look like spirals. Olfactory neurons sit near the surface of the nasal mucosa, where they can detect compounds that bind with their receptors. Underneath this mucous layer, Bang revealed thickly branched olfactory nerves, which transmit stimuli from the *olfactory receptor* neurons to the brain. Bang argued that all the anatomical evidence overwhelmingly suggested that these birds had a sense of smell. She later published, in collaboration with neuropsychiatrist Stanley Cobb, a comparative study of the olfactory bulb in the brains of 108 species of birds—this 1968 study is still frequently referenced today. Ironically, some textbooks have referenced this study's descriptions of small olfactory bulbs to back up claims that birds *can't* smell, which was far from the intent of the work.

In 1985, Bang and Wenzel teamed up to write a comprehensive summary of what was then known about avian olfaction, published as a chapter in the book series *Form and Function in Birds*. Their review covered olfactory bulb sizes, the surface area of the olfactory mucous membrane that lines the nasal cavities, the nerve connections between this membrane and the brain, and sensitivity to odors in the lab. At the time, few behavioral studies had been conducted in the field (excluding, of course, the very poorly designed experiments of the previous century), and Bang and Wenzel emphasized that more behavioral studies were badly needed to understand how birds used olfaction. Though their pioneering work was overlooked by animal behavior textbook writers for decades, it has been extremely influential to more recent studies of avian chemical communication.

MOM CAN'T SMELL

The scientific myth that birds can't smell has also crossed over into popular culture. Like many people my age, I was warned as a child that you should never touch a baby bird because, if you do, when the mother bird comes back, she will smell your scent and abandon the nest. But as a bird biologist, I learned that this notion just isn't true: I handled baby birds in the nest almost every day when I was doing fieldwork. As part of standard procedure, we weighed, measured, banded, and took blood samples from every nestling we found in our study site. If we were lucky enough to find the nest before the eggs hatched, we could collect data on the day the baby birds hatched, and every three days after as they grew up. We did not want to alert predators to the presence of the nest, so we were careful not to disturb the vegetation, keeping the nest hidden and avoiding creating an obvious trail by approaching the nest from different directions on each visit. Although we handled these nestlings every few days, parental abandonment of the nest was never a problem.

While I was drafting this chapter in August 2019, I conducted an internet search for "if you touch a baby bird, will the mom abandon it?" All of the top ten results stated that birds have a limited sense of smell and won't be able to detect your scent. These results come from respected and popular scientific sources such as the *Scientific American*, *National Geographic*, and even the Cornell Lab of Ornithology. Although some of the pages were several years old, one of them had been published just four months prior. I had

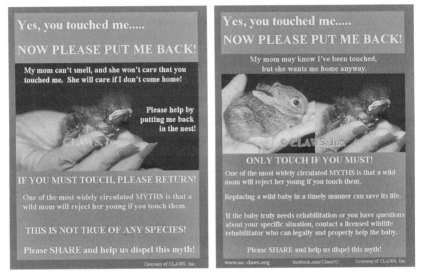

Flyers by the North Carolina wildlife rescue group CLAWS, Inc. Flyer on the left is the original version stating that birds do not have a sense of smell; flyer on the right is the new and improved (and scientifically accurate) version. *Images courtesy of CLAWS, Inc., https://www.nc-claws.org/index.php/rehabilitation/flyers*

found the same kind of faulty reassurance online a decade earlier when I first started trying to understand why people thought birds didn't have a sense of smell. Back then, I found a flyer from a North Carolina wildlife rescue group showing a picture of a hand holding a baby bird; the flyer read, "Yes, you touched me . . . Now please put me back! My mom can't smell, and she won't care that you touched me. She will care if I don't come home!"* Although I appreciate the sentiment (the urgent plea to "put the bird back in the nest!"), the science was wrong.

Another fun internet "fact" about bird smell is the uncredited assertion that great horned owls are the only natural predator of striped skunks because they have no sense of smell, and can

---

* Happily, this flyer has since been updated. It now shows a photo of a baby rabbit next to the baby bird photo and reads, "Yes, you touched me . . . Now please put me back! My mom may know I've been touched, but she wants me home anyway."

therefore chow down, blissfully ignorant of their meal's eye-watering miasma. In fact, there are several other animals that prey on skunks, including coyotes and bobcats—and no one doubts the olfactory capabilities of these mammals. Most of the time, the skunk's noxious spray does a pretty good job of deterring mammalian predators. But that doesn't mean owls and other raptors that eat skunks are anosmic.

My husband, Nathan Burroughs, is not a biologist, but like me, he loves animals. He also loves history. He had always been curious about falconry, the art of keeping and training birds of prey, as it brings together these two interests. A few years ago, he attended a weekend workshop put on by the Michigan Hawking Club to meet some active falconers and learn more about what their work involves. During one of the discussions, Nathan asked the instructors if the hawks used scent to find their prey. One of the instructors dismissed the idea, insisting that birds don't have a sense of smell; instead, raptors rely on their exceptional eyesight.

The instructor made this declaration while standing in front of a picnic table spread with the sandwich components for lunch, including the usual meats and cheeses. Also situated on the table was one of the demonstration birds, a merlin, hooded and tied to his perch. For an effectively blinded animal that supposedly had no sense of smell, it appeared to be working awfully hard to get loose from its perch and closer to some roast beef.

WHO ELSE CAN'T SMELL?

There is another group of animals* with an unfairly maligned sense of smell: humans.

---

* I should clarify that I am referring only to terrestrial animals. Cetaceans, the group of marine mammals including whales, are another group in which olfactory abilities have been doubted because of their greatly reduced olfactory anatomy, especially the toothed whales (including dolphins), but more recent research suggests that they may indeed have chemosensory abilities.

Just as in birds, this belief is not the result of any experimental study of human olfactory capabilities. Instead, we can trace it to nineteenth-century brain anatomist Paul Broca, best known for his discovery of the area of the brain involved in language production (now called Broca's area). He observed that the frontal lobe is much larger in humans than other mammals, and he posited that this part of the brain was the physical location of the "enlightened intelligence" unique to humans. In contrast, the olfactory region of the human brain is small. Plus, humans do not seem to display "odor-compelled behavior" like other mammals, so Broca deduced that the human sense of smell had atrophied. Thus, he concluded that intelligence and free will, not sense of smell, guided the life of humans and other primates.

In the early twentieth century, psychologist Sigmund Freud perpetuated this idea and incorporated it into his theories of psychosocial development. He insisted that the stages of early childhood focused on smell (as well as taste and touch) were merely playing out early forms of animal life. The human sense of smell remained untested for decades, with textbooks citing human olfactory organs as "vestigial." Even as late as 1958 a textbook about human olfaction opened with the sentence, "The human mind is an inadequate agent with which to study olfaction, for the reason that in Man the sense of smell is relatively feeble and not of great significance."

We commonly contrast our own, supposedly deficient, sense of smell with that of domestic dogs. Dogs can track people and animals, detect drugs and illnesses, and find their way home over long unfamiliar distances, all using their excellent sense of smell. Yet, humans are actually capable of some of the same olfactory feats that we admire in dogs. In a study at UC Berkeley, human volunteers were asked to follow a thirty-three-foot trail of chocolate essential oil through the grass, using only their sense of smell to guide them. The volunteers wore blindfolds, earmuffs, heavy gloves, and kneepads to ensure that they could not use sight, hearing, or touch during the trials (and, presumably, they were also asked not to taste the trail—the paper didn't mention anyone licking the grass). The humans were able to successfully follow the trail, and they got better and faster at it with practice. The human subjects even showed the same behaviors as a dog following a scent trail, including zigzagging across the trail to localize the scent and increased sniffing frequency as they moved faster.

Some humans can also detect illness using scent—within just hours of infection. An extreme example of this phenomenon is a British woman named Joy Milne, a "super smeller" who has the ability to smell Parkinson's disease. Many years before her husband was diagnosed with the disease, Milne had noticed that he had developed a "musky" smell. Later, when attending a patient support group with her husband, she realized everyone in the room had the same smell. Since then, Milne has made her unusual ability available to researchers looking for new methods of early detection of Parkinson's. Recently, studying the odor of Parkinson's patients led to the discovery that these patients' skin oil gives off a distinct odor signature. This discovery could lead to a new noninvasive screening procedure.

The role of smell in human social life isn't very well understood, probably as a consequence of scientists ignoring it for so long. We are beginning to learn that humans can distinguish between relatives and nonrelatives based on scent, and that we can pick up on all kinds of other cues as well. We pay attention to smell without knowing it. For instance, we unconsciously smell our hands after shaking hands with strangers. Exposure to the body odor of an ovulating woman can increase a man's *testosterone* levels, and men rate the scent of ovulating women as more attractive than the scent of nonovulating women.

The connections between unconscious odor preferences and sexual attraction may be unsettling, because they imply that we have less control over our mate choice than we believe. We enjoy thinking of ourselves as rational beings, not influenced by something so primal as scent. And we devote much energy to covering up our natural scents with soap and deodorant and perfumes, often finding natural body odor to be repulsive.

For humans, as for birds, people are quick to conclude that we could lose an entire sense just because we don't seem to "need" it. But I have the same fundamental objection to this assumption about humans as about birds. Natural selection doesn't work that way—there would have to be a disadvantage to retaining the trait, something that reduced one's survival or reproduction, in order for its loss to be an advantage. The only evolutionary examples I know of in which animals have truly, completely lost a sense they didn't need are cave-dwelling animals that live their entire lives in complete darkness, such as the Mexican tetra, also known as the blind cavefish. In this unusual situation, the energy and resources required to build a developing animal's visual system could be better redirected elsewhere, and animals who did so had an advantage. Interestingly, populations of Mexican tetras that are found in surface rivers have eyes, while those found deep in caves are eyeless. By comparing energy use in these populations, researchers were able to show that the young developing fish in the populations with eyes used about 15% more energy than the eyeless fish. Clearly, keeping a sense that you have no use for can be costly, and there are rare cases where you'd be better off to get rid of it. But while there are places on earth that are completely dark, there are no environments on our planet in which there is nothing to smell, taste, touch, or hear. It is difficult to imagine how it would ever be advantageous to lose any of those senses.

PHEROMONES AND OTHER SEMIOCHEMICALS

All living things produce chemical substances that other living things detect with the so-called chemical senses, smell and taste. Any kind of compound involved in the chemical interaction between organisms is called a *semiochemical* (from the Greek *semeion*, "mark" or "signal"). There are several classification systems for naming and thinking about different kinds of semiochemicals. Categories include: who is interacting (same or different species); what kind of information is conveyed (individual-specific or a species-wide signal); and whether the substance evolved for the specific purpose of communication.

The word "pheromone" is popularly used to mean all scents involved in attraction or sexual behavior, but the scientific term has a very specific meaning. A *pheromone* is a semiochemical that evokes a specific reaction in the recipient. For example, in nearly all species of moths, females emit a pheromone blend that attracts males. Male antennae have specific receptors tuned to the pheromones of their own species, and when these receptors are activated, the males start moving in the direction of the pheromone. These pheromone blends are species-specific, so they only attract members of the same species. However, *within* the species, they are essentially anonymous: they are exactly the same for all members of the same species, *conveying no individual information about the sender*.

Not all pheromones are sexual: in another well-documented example, rabbit milk gives off an odorous compound that induces suckling behavior in newborn bunnies. Again, this compound is the same in all rabbit milk: 2-methylbut-2-enal, or 2MB2 for short. Without any variation, it cannot convey any additional information about the mother.

Unlike pheromones, *signature mixtures* are semiochemicals that contain information about the individual sender. This information is present because the signature mixture is influenced by aspects of that individual's biology, such as the sender's genetic background, including MHC genes, as well as hormone levels, infection status, and even the bacteria the sender carries. These semiochemicals are the ones that hold the most interest for me, as they are involved in much more nuanced interactions than pheromones.

I like to think of the difference between pheromones and signature mixtures by comparing them to verbal communication. A pheromone is like a command. The female moth pheromone simply says: GET OVER HERE NOW. There's no additional information offered and no response other than to obey the command. Signature mixtures, on the other hand, can be exchanged more like a conversation (or an online dating profile).

My name is Sally. I'm an adult female with no offspring. I'm not currently ovulating, and I don't have malaria.

I'm Harry. I'm male, and I'm unrelated to you. I'm pretty good at fighting but unlikely to provide care for any offspring that I sire.

Another way to think about semiochemicals is to distinguish whether the semiochemical's primary purpose is for communication, or for something else. Indeed, animals can detect information about each other by observing many features that have not evolved for the purpose of communication. Such features are known as *cues*. A *chemical cue* is any compound or blend of compounds that contains information used by a receiver. For example, a bacterial or viral infection can change an animal's scent because it activates the animal's immune system. Immune functions produce chemical by-products that are expelled through the skin, breath, and other secretions and excretions, changing an animal's odor. Other animals may notice the smell and use that information to change their behavior toward the sick animal. If a fertile female is on the market for a mate, she is likely to avoid a male that smells sick. However, if she encounters a sick-smelling female that is one of her competitors for food, mates, or dominance rank, she may decide that now is a good time to attack.

In contrast to a cue, a *signal* is a trait (such as a smell, a song, or a feather color) that contains information about the sender and has evolved through natural selection because of its benefits for both the sender and the receiver. *Chemical signals*, including pheromones, can evolve from cues if they provide an evolutionary advantage. The peacock's tail is the result of countless generations of females preferring to mate with males with bigger, brighter, more elaborate tails, so the tail feathers got bigger and brighter and more elaborate over time. In the same way, a chemical signal can evolve if a chemical cue is attractive and if animals that produce more of the chemical have more offspring. Their descendants may also produce increasing amounts of that chemical, and eventually maybe even limit its production to breeding season, when it will have the greatest impact without wasting energy. Although we can envision scenarios in which these evolutionary advantages can occur, it's difficult to show that they have occurred without a time machine. So, many researchers—including me—avoid using the term "signal" and simply refer to most information-containing odors as chemical cues.

WELL, ACTUALLY . . .

In 2008, when Jim Goodson casually informed me that birds can't smell, there were multiple groups of chemical ecology* researchers actively publishing evidence to the contrary. At the time, most of this work, like that of Bernice Wenzel, focused on the tubenosed seabirds like albatrosses, petrels, and shearwaters.

Although scientists had known for decades that seabirds were attracted to food odors at sea, the exact molecules that enticed the birds had not been identified. In the 1990s, UC Davis researcher Dr. Gabrielle Nevitt, inspired by Wenzel, speculated that the birds were attracted to dimethyl sulfide (DMS), a small volatile compound given off by patches of phytoplankton in the ocean as a by-product of their metabolic processes. DMS smells a bit like rotting seaweed. Fish and squid feed on the phytoplankton, so the presence of DMS is a reliable cue to the birds that their favorite foods will

---

* Chemical ecology is the interdisciplinary study of chemically mediated interactions between living organisms.

Crested auklets performing a ruff sniff. *Photo by Ian L. Jones*

Several crested auklets in a scrum, all nape sniffing at once. *Photo by Julie C. Hagelin*

be found there. Nevitt found that several species of petrels at sea were attracted to vegetable-oil slicks with DMS added. Since DMS is airborne, she also tested whether the petrels were attracted to DMS-containing aerosols, and she found that the same species drawn to the oil slicks also showed up for the aerosols. Finally, to put it all in a natural context, Nevitt measured atmospheric levels of DMS at sea and concluded that seabirds could be found where DMS was highest.

From my perspective, the most exciting studies were those of birds using odors in social and reproductive behavior. The crested auklet, a highly social Arctic seabird with an adorable curly top knot, smells very strongly like tangerines* in the breeding season. Dr. Julie Hagelin, now a researcher at University of Alaska Fairbanks, discovered that the auklets were attracted to the citrusy scent. This smell, produced seasonally, seems to be particularly important in auklet courtship behavior: the birds rub their beaks in their partner's neck feathers, where the odor is strongest. Scientists call this behavior a *ruff sniff*.

---

* Fun fact: Although crested auklets smell like tangerines, the odor they produce contains none of the same compounds given off by tangerines. Instead, their odor includes 8- to 10-carbon aldehydes and alcohols (such as cis-4-decenal and octanol) that humans use to create artificial tangerine odor.

French researcher Dr. Francesco Bonadonna studies monogamous seabird species that live in massive breeding colonies, with thousands of birds building nests or burrows just inches apart from each other. He is interested in how they are able to recognize their partner and identify their nest among the densely packed colony. In particular, he wondered whether nocturnal burrowing birds like storm petrels, who couldn't rely on sight to recognize their burrow, use smell to find the right home. He captured several species of Antarctic seabirds, plugged their nostrils, and released them, testing how long it took them to find their burrow, if it all. Nocturnal burrowing species with plugged nostrils performed very poorly, while diurnal birds and surface nesters found their homes easily. Bonadonna later found through behavioral trials that the burrowing species preferred the scent of their own nest and their own breeding partner compared to random nests or birds from the same colony. These studies were among the first to suggest that odor could be important in avian social behavior.

The evidence seemed to be stacking up against the general consensus that birds had no sense of smell, given the findings of these seabird studies. But what about songbirds (also called passerines), the largest group of birds in the world, which have the smallest olfactory bulbs of all? The very year that I began thinking about these questions (2008), two studies had been published that showed songbirds could detect the scent of mammalian predators. In one, house finches were hesitant to approach bird feeders near the scent of cat feces. In the other, blue tits were much more vigilant around their nest boxes when the scent of a ferret was present.

No one had yet investigated whether songbirds used odor to communicate with each other, but it seemed to me that there was no reason that they couldn't. That same year, Dr. Silke Steiger at the Max Planck Institute for Ornithology sequenced olfactory receptor (OR) genes in nine bird species. Animal genomes have many OR genes, often hundreds, all slightly different, because specific

OR proteins bind to specific odors in the environment, requiring different genes for each of the receptor types. For example, the characteristic scent of vanilla comes from a molecule called vanillin. To detect the scent of a baking cake, you would need genes for the receptor that binds to vanillin, as well as receptors for other molecules given off by the eggs, sugar, and flour in the cake. To get a sense of an animal's olfactory ability, we can look at the total number of functional OR genes compared to the number of *pseudogenes*—genes that in the evolutionary past were functional but no longer produce the proper protein sequences due to mutations somewhere along the line. All of the birds in Steiger's study had high numbers of functional OR genes—and the small songbirds she included, the blue tit and the canary, had just as many genes as the snow petrel, a seabird with the largest olfactory bulb of any bird. There was also behavioral evidence that songbirds could smell: fifteen years earlier, researcher Dr. Larry Clark at the Monell Chemical Senses Center in Philadelphia had trained five species of songbirds (gray catbirds, eastern phoebes, black-capped chickadees, great tits, and European goldfinches) to give a conditioned response to odors. The birds were first trained by pairing odors with a small electric shock delivered to the leg. In later trials, the birds' heart rates increased in response to the same odor presented without the accompanying shock. The odor detection threshold (the ability to detect very small amounts of an airborne compound) of these birds that were supposedly *microsmatic*—meaning they have a poor sense of smell—was comparable to that of *macrosmatic* species—those with a strongly developed sense of smell—such as rabbits and rats. Science had clearly underestimated the olfactory abilities of songbirds. These intriguing studies captivated me, and I decided this seemed like an excellent time to add a new dimension to my research on dark-eyed juncos.

CHAPTER 2

FOLLOWING THE
BIRD'S NOSE

There is nothing of especial interest in the Junco's habits, and only a bird-lover can imagine what a difference his presence makes in a winter landscape.
　—FRANK M. CHAPMAN, *Bird Life:*
　　A Guide to the Study of Our Common Birds

50% of fieldwork is sheer terror. The other 50% is boredom.
　—JOEL MCGLOTHLIN, Ketterson lab alum,
　　now associate professor at Virginia Tech

WHO CARES ABOUT JUNCOS?

"Why do you study dark-eyed juncos? Why don't you pick a more interesting bird?"

I get this question a lot. Dark-eyed juncos (*Junco hyemalis*, the species name comes from the Latin for "of the winter") are small gray and white sparrows that are common throughout much of North America. Although they are cute and birders enjoy seeing them, they don't have colorful feathers, they don't have fancy courtship dances, and their song is a rather unremarkable trill. They don't build complicated nests or bowers, eat anything

Banded female dark-eyed junco at Mountain
Lake Biological Station. *Photo by Nicole Gerlach*

unusual, or live in extreme environments. So, what's so special
about juncos?

To be blunt, the answer is . . . nothing. These normal little birds do
normal little things—and that's the brilliance of focusing on them
as study subjects. They are easy to find in large numbers throughout North America in various habitats and conditions. Conclusions from our research can help us understand not just juncos
but most birds with similar biology, like the many species of sparrows or other small seed-eating birds like cardinals, titmice, and
chickadees. Studying something like, say, the elaborate mating
rituals of the satin bowerbird in Australia is extremely interesting
in its own right, but you may not be able to apply your findings to
other animals.

A question I've been asked: "Why don't you study something
exciting, like bald eagles?"

My response: "Have you ever tried to catch a bald eagle?"

Juncos are pretty simple to observe and to capture. They are
mostly ground feeders, eating seeds and insects. They also nest on
the ground, typically hiding their open-cup nests underneath some

overhanging foliage, sometimes on a slight slope. The fact that they spend so much time on the ground means we're more likely to see them, unlike birds that are high in the forest canopy. Because we can attract juncos with piles of seed as bait, they are relatively easy to catch in mist nets.* They also do well in captivity with this uncomplicated diet, often surviving much longer than they do in the wild, and sometimes even reproducing. (It can be difficult to provide the proper conditions for a wild animal to breed in captivity, but we try.)

Juncos behave like most temperate Northern Hemisphere songbirds, migrating in response to seasonal weather patterns and shifting their social structures accordingly. Adapted to cool climates, juncos are often called "snowbirds" as they spend their winters in mixed-sex flocks that forage together and are frequently seen at bird feeders during the snowy months throughout the US and in southern parts of Canada. During the summer, they move back to cool-climate breeding ranges, either much farther north in Canada or to higher altitudes in the Rockies and Appalachians. There, they split off into territorial male and female pairs. These pairs work together to raise their offspring and defend their territory from other juncos. This social system, found in over 90% of birds, is what we call social monogamy: the male and female are pair bonded, at least for the season. Their social behavior reflects that bond; they spend most of their time near each other, and they will fight any would-be suitors that show up. Until the age of paternity testing, researchers assumed that socially monogamous birds were also genetically monogamous, meaning that they were sexually faithful to each other and did not mate with birds outside

---

* Mist nets are made of fine-mesh nylon or polyester panels that are difficult to see in sunlight, usually a couple of yards high and stretched between two poles twenty to forty feet apart. The nets' horizontal lines create large loose pockets that entangle flying birds. In addition to birds, we sometimes catch small mammals, such as chipmunks or bats, as well as our own watches and zippers. Once, we found part of a sandwich in our net, probably tossed out a car window.

of their bond. But thanks to modern technology, we now know that *genetic monogamy* is incredibly rare in any animal. In all of the junco populations that have been studied, paternity tests show that about one-quarter to one-third of all junco nestlings are *extra-pair offspring*—that is, they were fathered by a male other than the one that raised them. (Nestlings sired by both of their social parents are called *within-pair offspring*.) This lack of faithfulness is extremely common across all birds,* again making the junco quite typical.

So, why study such a normal bird? Well, I find that normal is only boring if you think you already know everything about it. And I think we are far from understanding the reasons why "normal" is normal in the first place.

FOLLOWING MY OWN NOSE

While I didn't always study juncos, or even ornithology, I have been interested in mate choice for a long time. During graduate school at CUNY, I thought a lot about how monogamous primates, such as gibbons, titi monkeys, and some humans, select their mates. Choosing just one individual to pass on genes to all of your offspring seems quite a high-risk proposition. Indeed, monogamy is rare in primates and, more broadly, in mammals—only about 3% to 5% of mammal species are monogamous. Many monkeys and apes live in *polygynous* (from the Greek *poly*, "many," and *gyne*, "wife") "harem" groups with one dominant male and several females that mate with him, as in baboons or gorillas. Others, such as chimpanzees and squirrel monkeys, live in even larger groups with multiple males and multiple females, and more complex mating structures. Still others, such as some tamarins and marmosets, have highly specialized reproductive systems in which the female always gives birth to twins, necessitating multiple male partners to help care for them.

* And humans!

The long-armed, tree-swinging gibbons I studied in Southeast Asia, along with their larger cousins, siamangs, are the only "small apes," compared to the "great apes" group that includes chimpanzees, gorillas, and orangutans. Gibbons and siamangs are also the only apes that are monogamous. In my original research proposal, I wanted to know how young adult Kloss's gibbons, a rarely studied species, set out from their parents' territory and found other, unrelated, gibbons to be their partners. To learn such details about an animal's life and behavior, you need to see what they are doing, and it helps to get DNA samples for genetic paternity and relatedness testing. Unfortunately, most gibbon species are endangered, and they live in remote rain forests, so they are difficult to find and follow. Further complicating matters, catching them to take samples is generally not allowed. Instead, you have to rely on noninvasive samples, like poop. Fecal samples contain cells shed from the lining of the animal's intestines, so it's possible to get the animal's DNA if you can be sure to get the outermost part of the dropping. That's easy when you're studying, say, elephants, who leave giant piles on the ground. But when you're dealing with an animal that lives high in the trees, things can be much trickier.

In 2001 and 2003, I spent a total of ten months trying to collect fecal samples from Kloss's gibbons in the Mentawai Islands of Indonesia. Although you can hear their beautiful calls from more than half a mile away, the gibbons spend most of their time in the upper reaches of the forest canopy—up to 130 feet off the ground!—and are very difficult to see. When they defecate, the seed-filled feces (they eat mainly fruit) shatter into small fragments all over the forest floor. Getting them to poop wasn't too difficult. My sample collection method was as follows: (1) follow the gibbon's song until you are able to glimpse the group, and (2) wait nearby until they see you. Inevitably, once the gibbons spotted me, they would all give alarm calls and run away—some of them leaving samples in the process.* The next step, (3) find the fragmented samples, was much more challenging. In addition to the difficulties inherent to picking out seedy poop shards from the forest floor, sometimes I would find feces that came from different animals entirely.

There were four different species of primates in the Mentawai Islands. In addition to the Kloss's gibbons (Hylobates klossii), there were Mentawai macaques (Macaca pagensis), Mentawai langurs (Presbytis potenziani), and simakobu monkeys (Simias concolor). Langurs and simakobus are folivorous, meaning their diet consists primarily of leaves, while macaques and gibbons mostly eat fruit. Dietary and other factors often caused the various species to leave distinctive fecal samples. In retrospect, smell played an important role in this research project, too; I became quite adept at telling the difference between the poop of these four primates on the basis of smell alone.

After nearly a year in the rain forest, I had a grand total of thirty-one Kloss's gibbon fecal samples to show for my efforts. I spent about two more years trying to extract gibbon DNA from these samples and was only successful with twenty-one of them. Given the difficulties I had observing gibbon behavior and obtaining genetic data, the focus of my dissertation research shifted to conservation biology out of necessity: understanding how these gibbons were related to other species (their closest genetic relative is the Javan gibbon), whether there were different subspecies of Kloss's gibbons (no), and just how many of them were left (not many—my best estimate was twenty thousand to twenty-five thousand in 2005).

I have the greatest respect for conservation biologists. I care very much about conserving the rain forest and the wildlife in Indonesia, but I also found it disheartening. It often feels like you are fighting a losing battle, especially in areas where people depend

* I called this the "scream, shit, and run" method.

so heavily on these natural resources for their own survival. After graduation, I decided to return to the original behavioral questions that motivated me.

Although monogamy—both social and genetic—is rare in mammals, social monogamy is the norm in birds. Plus, birds are everywhere. I figured that if I turned my attention to studying our feathered friends, I wouldn't have to spend months on end trying to secure research permits and travel visas from foreign governments. I wouldn't even have to risk getting bitten by leeches (a constant problem in the Mentawais*). Birds seemed like the perfect choice for my next act.

But I didn't know anyone who studied birds. My PhD was in an anthropology department, without many links to researchers in biology departments. Serendipitously, while applying for dozens of academic jobs, I stumbled across an advertisement for a position managing Dr. Ellen Ketterson's laboratory at Indiana University. The ad described Ketterson's long-term project on dark-eyed juncos. Eureka! Birds!

At the time, her lab primarily focused on endocrinology methods like *hormone assays* (a method to measure how much of a hormone is present in blood or other types of biological samples), because they were interested in how testosterone levels influenced behavior. I had no experience with either birds or hormone assays. But I had spent the last several years developing DNA *sequencing* and genotyping skills, which the Ketterson lab was just starting to use. I hoped that my expertise with fieldwork and genetic work would be seen as beneficial enough to excuse my lack of experience in ornithology and endocrinology. I submitted my application but heard nothing back.

After a while, I did something that was a bit terrifying at the time. Of the dozens of academic positions I had applied to, this felt like

---

* My husband asked me to also mention malaria. He got malaria when he came to the Mentawais with me, and according to him, I can't talk about the Mentawais without mentioning that he got malaria and I didn't.

the right one, so I tried harder. I wrote to Dr. Ketterson again to clarify why I was so interested in the job and why I would be a good fit, even though on paper I seemed completely wrong for it. I described why I wanted to work with birds instead of primates. I explained that I had years of fieldwork experience in challenging environments and could easily learn ornithological methods. I listed my laboratory expertise and elaborated on how beneficial it could be to her research group, and how easily I could learn to do hormone assays and why they were important for my research too.

She wrote me back. I got the job.

I loved it, right from the beginning. True, I had a lot of catching up to do—as it turns out, birds are *really* different from mammals. But it was refreshing and fun to be doing something so new after over seven years working on just one question. I was going to study how a bird's genes led it to choose another bird as its mate. I planned to unlock nature's secrets of perfect pairings.

At least, that's what I thought I was going to do. A recurring theme in my research, and my life, is that things rarely go as planned.

THE LONG-TERM JUNCO STUDY

Dr. Ellen Ketterson is a small, energetic woman with spiky hair and a quick smile. She handles birds with ease and precision—it's obvious that she cares deeply about their welfare. I was initially surprised to learn that she had once been a cheerleader in high school, but once I experienced her positivity and relentless support of her students and colleagues, it made perfect sense. She appreciates adventure and has traveled the world to study birds in places like the cloud forests of Costa Rica. In addition to these enviable traits, Ellen Ketterson is a dedicated scientist who speaks with authority gained through decades of research and expertise.

Ketterson began studying juncos at Mountain Lake Biological Station in 1983. She has been running a long-term study ever since, with the help of generations of undergrads, graduate students, and postdocs.* Mountain Lake Biological Station, run by the University of Virginia, is situated in the Appalachian Mountains in southwestern Virginia, just a couple of miles away from the Appalachian Trail. It's about four thousand feet in elevation and is surrounded by forests that stay cool in the summer and are often misty in the mornings. The habitat is ideal for dark-eyed juncos, which thrive at high elevation and high latitudes. It's also an ideal habitat for researchers and students, with lab space, cabins, a small cafeteria, and even a pond for swimming.

Long before I knew anything about juncos, Ketterson was asking many of the same questions that I was interested in. Especially, why were these birds monogamous? Sex hormones, like testosterone and estrogen, were one possibility. Biologist Dr. John Wingfield compared patterns of testosterone levels across various species of birds with different mating systems. In polygynous species with more aggressive and less paternal males, the males have high levels of testosterone circulating in their bloodstream throughout the length of the breeding season. In contrast, males in monogamous species, who typically exhibit more parental care and less aggression, have testosterone levels that only briefly peak at the beginning of the breeding season—when they are courting females and defending territories—and then drop almost to non-breeding-season levels while they are caring for their nestlings. Perhaps the best answer to the question *why be monogamous?* is simply *because parents need to help take care of the kids.*

Around the same time as Wingfield's study, Ketterson was studying parental care in juncos at Mountain Lake. She removed mated males from their nests and measured the effects on their partners and offspring. Females whose mates had been taken away worked harder to feed their nestlings. Even so, only about half as many

---

* You can watch a great documentary about Ketterson's work, *The Ordinary Extraordinary Junco*, at http://juncoproject.org.

of their offspring survived, compared to females who had mates helping them. While the lack of help was clearly bad for the females, these results suggested that males could, as an alternate strategy, successfully produce the same number of offspring if they mated with at least two females and didn't spend any energy helping them. Why then did males invest the energy in staying with and helping one female raise the kids? Maybe there were other benefits to mating monogamously—or perhaps there were costs to *not* being monogamous.

Ketterson approached this question with a method she coined "phenotypic engineering," in which scientists produce the traits (*phenotypes*) that they want to see in an individual by manipulating its physiology. If males continued to seek mating opportunities throughout the entire mating season, rather than helping to raise their offspring, we would expect them to have continuously high levels of testosterone throughout the breeding season. What would be the downside? To answer this question, Ketterson created artificially high testosterone levels in male juncos by using testosterone implants: tiny pieces of silicone tubing packed with crystalline testosterone inserted just under the bird's skin. These implants work much like implanted hormonal birth control in humans, slowly and steadily releasing the hormone into the bird's bloodstream. As expected, males with artificially increased testosterone spent less time feeding their nestlings and defending the nest. They also sang more and courted females more, making them more attractive to females. High-testosterone males had larger home territories and sought out mating opportunities outside their pair bond more often. They also sired more extra-pair offspring.

However, this philandering came at a heavy cost: high-testosterone birds had lower survival rates. They had heightened levels of *corticosterone* (a hormone secreted by the adrenal gland and associated with the stress response, similar to cortisol in humans),

reduced immune function, and decreased self-maintenance behaviors like preening. Their annual molt, in which birds grow the
fresh new feathers so important for winter migration, was even
inhibited. So, high levels of testosterone increased short-term reproductive success for males at the expense of decreasing their
survival and likelihood of future reproduction—which could be a
very big cost in a bird that can have five or more years of successful
reproduction. Was this trade-off enough to account for the testosterone patterns in monogamous birds, or was there more at play?

Ketterson suggested that while certain traits may be beneficial to
a male and thus passed on from father to son, those same traits may
not be as beneficial to daughters. Evolutionary forces can be very
different for males and females, which can lead to a phenomenon
known as sexual conflict. With the exception of the sex *chromosomes*, males and females have all the same genetic material, and
they pass their genes on to their offspring of both sexes.

Females naturally produce testosterone, and like males, their
levels peak during reproduction. We think of testosterone as the
"male" hormone and estrogen and progesterone as the "female"
hormones, but these compounds are actually all structurally similar and part of the same steroid synthesis pathway. In female
birds, much of the testosterone is converted to estradiol through a
process called aromatization* by the enzyme aromatase. But some
testosterone remains unconverted and can have other effects on
female physiology and behavior. What would the consequences
be if females inherited naturally high levels of testosterone? To
answer this question, Ketterson also put testosterone implants in
female juncos at Mountain Lake.

Just like males, females with high testosterone spent less time
exhibiting parental behaviors like *brooding* and defending the nest

---

* Despite the name, in chemistry, aromaticity has nothing to do with smell.

from predators. In fact, they decreased all reproductive efforts. Females with high testosterone were less likely to develop a brood patch (a featherless, vascularized area on the chest that increases skin contact to eggs and nestlings to keep them warm) or build a nest. Even the females that did successfully build a nest delayed egg-laying until later in the season and laid fewer eggs overall. As a result, these females had fewer surviving offspring than control females (who had empty implants). The testosterone also negatively affected the health and survival of the females: like males, females with extra testosterone had higher levels of corticosterone, reduced immune function, and a delayed fall molt (which could reduce their ability to survive migration).

The Ketterson lab wanted to know whether high testosterone led to increased extra-pair copulations in females. In general, scientists think it makes perfect sense for males to mate with extra females because it increases the number of offspring they sire. But females are more restricted in the number of eggs they can lay in one season. Why mate with additional males, especially if your social mate is providing benefits like nest defense and parental care? As it turned out, females with artificially increased testosterone levels did not have any more extra-pair offspring than females with normal levels.

What does drive females to mate with extra males, then? Perhaps females have conflicting priorities when choosing a mate: the male that will do a good job taking care of offspring may not have the best quality genes to pass on to those offspring. One solution would be to choose a male with good parenting skills as your social mate, and then choose a different male, perhaps an aggressive one with more colorful feathers, to actually sire the nestlings. Females could also increase their chances of success by having multiple fathers for a clutch (the avian version of a litter). The group of offspring overall would have increased genetic diversity, so the chances of at least some of them surviving would be higher.

These questions and more were under investigation in the Ketterson lab when I joined. It was the perfect fit for someone who was interested in understanding how monogamous animals choose a mate. Plus, the wealth of background knowledge and foundational data already generated by the lab was incredibly helpful for someone who had absolutely no idea what they were doing.

LIFE ON THE JUNCO CREW

My favorite part of junco work at Mountain Lake Biological Station is the annual spring early season census. Beginning in April, the birds migrate back up the mountains to breed after wintering in the milder temperatures at lower elevation. The birds have not yet established territories, found mates, or started building nests, so it's easier to catch them at central locations around the study area. Every spring during the census, we put up dozens of mist nets and traps to catch as many adults as possible. We attach a metal band (from the US Fish & Wildlife Service) marked with a unique number to every captured junco; we also add a different combination of colored plastic bands to each bird that can be seen from distance with binoculars. Many of the birds are already banded from previous years; they also get captured and documented so that we know who survived the winter and returned to the site. Juncos make up the majority of the catch at Mountain Lake, since they are so abundant and are attracted to the seed bait spread on the ground. But a big part of the fun of early season is the bycatch, the other species that get caught in our nets. If you're capturing juncos at Mountain Lake, you can usually count on also catching chipping sparrows, northern cardinals, titmice, nuthatches, several kinds of warblers, blue jays, robins, the occasional woodpecker, and, if you're unlucky, rose-breasted grosbeaks, who have a really painful bite.*

* The first time I saw a rose-breasted grosbeak in one of our nets, I was really excited. Dawn O'Neal (then a Ketterson lab PhD student who was teaching me to band birds)

: 39 :

Although the birds can get quite tangled in the nets, they are very rarely hurt in the process. Everyone working with the birds is trained how to safely hold them and cautiously remove the netting a bit at a time, being careful not to injure the legs or wings. Non-juncos are released immediately. Each junco we catch is then put into a separate paper lunch bag for transportation back to the lab at the research station. The brown paper bags let in air, so the birds can breathe easily, but also keep out most of the light, so the birds stay calm. In the lab, we take a series of body measurements, including wing length, tail length, and mass, plus blood samples for DNA and hormone analyses. After all the data are collected, the birds are released back at the site of capture, free to carry on with their day. Juncos are easy to handle, and many of them don't even seem too fazed by the experience. In fact, there are always a few birds who show up in the nets several times each week, attracted by the millet seeds. Apparently, for these birds, the benefits of the free lunch outweigh the inconvenience of being captured and prodded.

Around mid-May, the weather gets warmer and sunnier, and we catch fewer juncos in the nets each day. At this point, most of the juncos have found mates and claimed territories, and they prefer to stay close to home. So we pack up the nets and the traps and turn our attention to nest searching and monitoring. Juncos build small open-cup nests made of twigs and straw, lined with moss and soft hair plucked from where shedding deer have brushed up against tree and bush branches. We often find nests by poking around in the types of vegetation where we know juncos like to nest. Sometimes, we discover a pair carrying nest material, and we watch them patiently with binoculars to see where they are taking it. Once we find a nest, we keep an eye on it to see if it stays

said, "Have fun with that. I'll wait in the car." I quickly learned why she was not happy to see the beautiful birds.

active—in other words, if the female successfully lays, incubates, and hatches eggs, and raises the nestlings without being detected and eaten by a predator, or without abandoning the nest for some other reason.

When a junco begins laying eggs, she will lay one a day, usually before dawn, until she has finished the clutch. Most junco clutches have four eggs. However, the mother does not settle down on the eggs until all of them are laid. This way, all of the embryos will develop and then hatch at the same time. She incubates the eggs for about twelve days, leaving the nest every hour or so for a short period to feed herself. When the eggs hatch, the nestlings are tiny, about half the size of your pinky finger, with closed eyes and almost no feathers. Both the adults bring food, mostly larval and mature insects, to the nestlings throughout the day. In between feeding bouts, the female sits on top of the nestlings (brooding) to protect and warm them while they continue to develop. The nestlings grow quickly during this period and will reach nearly adult size in just eleven or twelve days. We monitor the nest throughout this period, weighing and sampling the nestlings on days zero, three, and six. After day six, we monitor the nest from afar—at about day eight or nine, the nestlings will jump out of the nest if threatened. Although nine-day-old nestlings have a better chance of surviving outside the nest than if they sat still and watched a snake approach to eat them, their chances are much better when they are fully developed, at day eleven or twelve. So, we are careful to keep our distance until they're ready.

Fledging Day, the day the nestlings can leave their cozy nursery, is an exciting time for junco researchers. Although we can identify the parents if they are already banded, we don't always get the opportunity to sample them earlier in the season. It can be difficult to find both parents at the nest at the same time and to capture them without endangering the nestlings. Fledging Day is our best chance to safely handle the entire family all at once.

Capturing birds on Fledging Day is always dramatic. Before dawn, you hike out to the nest and stealthily open a mist net placed there the night before. You retreat and quietly sit, awaiting the right moment. Once it is light enough for you to see, but before the birds wake up on their own, you sneak up to the nest. The female, who has spent the night keeping the nestlings warm, will startle, and react by flying off the nest. Then, you must act quickly. As carefully as possible, you reach into the nest—in the words of Ketterson lab alum Dustin Reichard, you must "strike like a snake"—and try to grab all four nestlings at once. Remember, they are nearly adult sized at this stage. If you don't have large hands, grabbing them all at once can be very challenging, and one or more might get away from you, running around on the ground and shrieking. Then, you have to try to catch them—although, I believe that part is a perfect job for spry undergraduate field assistants. The nestlings don't fly yet, but they sure can run.

As stressful as catching the nestlings can be, it gets worse. The parents, who are both nearby, hear their distressed nestlings and fly at your head attempting to defend their offspring, shrieking even more piercingly than the nestlings. Now the open mist net becomes very important. If you're lucky, both parents will get themselves caught in the net. Then you can gently remove them from the net, pop them into their own paper bags, and while the sun rises, you quietly sit on the grass and sample and measure the whole family. When you're finished, you release the parents, carefully place all four nestlings back in the nest, and leave the area quietly, letting them get on with their birdy lives. The nestlings usually jump out of the nest on their own afterward, perhaps a few hours earlier than they would have without our intervention, but they are developmentally ready. That day, they begin the process of learning to fly, and their parents keep a close watch on them and continue feeding them for days, sometimes weeks.

Most of the time, Fledging Day goes smoothly. My least favorite part is usually getting up before dawn and hiking out to the nest in the dark. But on my worst Fledging Day I discovered that a predator had gotten to the nest before I did. On another occasion, the nestlings got away from me—not unusual, as I noted above, but that time I never did recapture the fourth one. That incident is forever recorded in the lab's nest logs, where I wrote: "Fledged 4 D11 [day eleven] nestlings. Only processed 3—one got away from us because DJW [that's me] is a terrible excuse for a biologist."

Excerpt from handwritten nest logs in which I confess my very painful mistake.

Occasional unfortunate blunders aside, we typically close out the fledging sitting in a circle on the grass, taking data, and enjoying the sunrise. When we're done, we pack up our data collection tools, roll up the net, and head back to the station feeling accomplished. Then, we enjoy a hot breakfast at the station dining hall—eggs, bacon, and potatoes or pancakes, whatever the cooks made that day. It might not sound like much, but a real breakfast is the ultimate luxury since junco researchers usually head into the field before the dining hall even opens, and we come back long after the last waffle disappears. After the (mostly) controlled chaos of a fledging, the rest of the day feels almost dreamy, leisurely and serene.

JUNCO SMELLS

In the summer of 2008, I planned to collect DNA samples from adult pairs to test whether they chose mates based on their compatibility at major histocompatibility complex (MHC) genes. But when it was brought to my attention that so many people thought birds couldn't smell, I realized that I first had to address this issue before I could delve more deeply into the genetics. After all, MHC genes are thought to be detected by odor.

Until that summer, I hadn't thought much about what juncos smelled like. I normally only noticed their odor when there were many of them in a closed space, like the aviary, and then it was mostly their feces that I smelled. But now I started to pay attention. To me, juncos smell a little musty and a bit like the forest—notes of soil and leaf litter. My friend Nicki Gerlach, another Ketterson lab alum, says that to her, they smell like "ferns and Celestial Seasonings' Country Peach Passion tea." I find it to be a subtle odor, but perhaps it is more meaningful to the juncos.

I decided to add a side project to my fieldwork. Fortunately, some background work on junco odors had already been conducted. One of the graduate students in the lab, Sara Schrock, had just finished a project with chemists Dr. Helena Soini and Dr. Milos Novotny from Indiana University's Institute for Pheromone Research, studying a secretion called *preen* oil.

Unlike mammals, which have several different kinds of glands on the surface of their skin that secrete oils and odors, birds really only have one: the *uropygial gland*, also called the preen gland, which is found only in birds. You can see the tip of this gland by moving aside the feathers at the base of a bird's tail. When a bird preens, it periodically reaches back and rubs this gland with its bill, collecting a bit of oil and then spreading the oil on its feathers. Preen oil helps protect the bird's feathers from the everyday wear and tear caused by exposure to the elements and by ectoparasites, such as feather mites and lice. Regular application of preen oil maintains the appearance of feathers, making their colors appear brighter and more attractive to potential mates. In birds that spend time in water, such as ducks and seabirds, preen oil is waxier and helps with waterproofing. And, last but definitely not least, preen oil is a source of body odors in birds.

A junco's uropygial gland

Eggs (and nestlings) make an excellent meal, so hiding nests from predators is an important part of ensuring offspring survival. Just a few years before my interest in bird smell began, Dutch researcher Dr. Jeroen Reneerkens was studying preen oil in the red knot, a sandpiper that breeds in the high Arctic and nests on the ground. The red knot and other sandpipers switch the composition of their preen oil from monoester waxes, which have a low molecular weight, to heavier diester waxes during courtship and incubation. Lighter compounds are more volatile and airborne and, thus, can be smelled more easily than heavier compounds. In this study, a trained German shepherd had a more difficult time tracking the scent of diester waxes. Reneerkens concluded that the birds changed their preen oil composition to make the nest less smelly and therefore less detectable by predators, a trick he called "olfactory crypsis."

Back in Indiana, Sara Schrock, along with Soini and Novotny, wondered whether juncos do the same thing. Like the sandpipers, juncos are ground nesters, and they are highly susceptible to predators like snakes and rodents. (After years of discovering that chipmunks had eaten the eggs—or even nestlings—that I was monitoring, I stopped thinking chipmunks were cute.) The researchers took preen oil from juncos in breeding condition and juncos in nonbreeding (winter) condition and measured the odorous compounds with a technique called *gas chromatography–mass spectrometry*, which helps detect and separate the parts of a gas.

Schrock and the chemists were surprised to find the opposite of what they expected: the preen oil of juncos in breeding condition produced *more* compounds than juncos in nonbreeding condition. This increase in odiferous compounds seemed maladaptive, as it would likely make it easier for predators to detect nesting birds. But my first thought upon learning about this result was that *odor must be important in breeding behavior*. Maybe songbirds used odors to attract mates, just like mammals and insects.

But before I could start chasing that tempting idea, I first had to find out whether juncos could smell preen oil at all. Since I was already catching birds during early season and monitoring junco nests anyway, I decided to try a little experiment with nesting females—the perfect captive audience, since females were stuck sitting on their nests for hours every day.

My plan was to apply preen oil to junco nests while the females were away from the nest and then see whether the females noticed the smell upon their return. To figure out whether they "noticed," I would measure whether there was any change in the time they spent sitting on their nests. If the smell of another bird was threatening or disturbing in any way, the female might leave the nest more quickly than if the odor wasn't there. I used three different experimental treatments. As a control, one group of the nests would receive preen oil previously taken from the incubating female herself, which we would not expect her to respond to. To a second group of the nests in the study, I would apply preen oil taken from other juncos—unfamiliar individuals from the captive colony back at Indiana University. Finally, the third group of nests would receive preen oil taken from a different species. I collected these samples from northern mockingbirds at another lab at IU.

Mockingbirds pose no threat to juncos, but juncos would not normally encounter them at Mountain Lake, so this would be a very unfamiliar bird odor. The plan was to record females sitting on their nest for four hours on one day, and then the next day, add the preen oil to the nest and record them for another four hours.

I spent several tedious weeks setting up video cameras a respectful distance from nests, and several even more tedious weeks of watching and scoring mostly very dull videos of birds sitting on nests, but the results were clear. As expected, when I added the female's own preen oil to the nest, she did not change her behavior. The female would spend about the same length of time sitting on the nest before going off to search for food compared to her average incubation bout from the previous day. However, if I applied the preen oil of an unfamiliar junco—or worse, a mockingbird— she would almost always get up and leave much more quickly than an average incubation bout.

In my favorite video, the female returns to her nest shortly after I had applied mockingbird preen oil. She settles in to incubate her eggs. Suddenly, she starts making jerky motions, looking all around her. It seems that she has smelled something she doesn't like! The female junco then starts pecking around at the edge of her nest, right where I applied the drops of preen oil. She picks up several pieces of twigs and other nesting material and flies away from the nest, taking the twigs with her.

These results seemed conclusive to me: the birds could definitely sense that I had added preen oil from other birds to their nests, and smell was the most likely explanation. It was also clear that they didn't like the scent of an intruder at their nests. Like any inveterate researcher, I immediately had more questions. Now I was intrigued not just by how birds might use odor in mate choice but also by the importance of scent in the nest itself.

In particular, I started thinking about brood parasites—birds that lay their eggs in other birds' nests. We often found eggs from

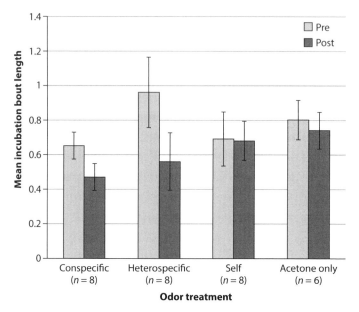

Average length of time female juncos incubated their nests before and after I added preen oil from conspecifics (other juncos), heterospecifics (mockingbirds), self (the female's own preen oil), or acetone with no preen oil. Females whose nests had been treated with preen oil from other birds reacted by spending less time on their nest than they had before treatment. The number of birds in each group is represented by n. *Adapted from Whittaker et al. 2009, Journal of Avian Biology 40:579–583*

one such brood parasite, the brown-headed cowbird, in junco nests at Mountain Lake. If we did not remove those eggs ourselves, the juncos did not seem to notice that one or more of the eggs they were sitting on were not their own. One day, while removing one of these cowbird eggs, I began to wonder how they got away with it. If juncos quickly noticed the smell of another bird in their nest, why didn't they do something about it when a cowbird left such obvious evidence?

FRESHLY BAKED COWBIRDS

Brown-headed cowbirds, Molothrus ater ("molothrus," from the Ancient Greek words for "struggle" and "impregnate"), belong to the blackbird family Icteridae. Like other brood parasites, they don't build their own nests or raise their own offspring.* Instead, they lay their eggs in the nests of other species (called hosts), usually one egg per host nest, mixed in with the host's own eggs. The victim will unwittingly incubate the cowbird egg along with her own. Cowbirds are known to parasitize the nests of over 220 species, with at least 144 of those species observed to have successfully raised cowbird nestlings.

Most host species are much smaller than cowbirds. An adult brown-headed cowbird is about 1.4 to 1.8 ounces; by comparison, juncos are only around 0.7 ounces. In most host nests, the hatched cowbird nestling looks different from the resident nestlings, yet the host still feeds and raises the cowbird. Caring for the cowbird nestling comes at the expense of the host's own nestlings, particularly in smaller host species, as the feeding rate is always higher for the cowbird nestling. The biological urge to feed a begging nestling must be incredibly strong if the feeding continues after they've grown larger than the host parents themselves.

If, as I had already shown, juncos noticed the smell of a few drops of preen oil applied to their nest, surely they should recognize the scent of another species laying a whole egg in it. I wondered if cowbirds were odorless, or if maybe their scent blended in with the nest material or other things in the environment that the host bird would not find unusual. I immediately wanted to know what cowbirds smelled like, so after getting the appropriate permissions, I set up some baited nets and traps in an open field frequented by cowbird groups. When I caught my first couple of cowbirds, I bundled them up in paper bags, just like we do with juncos, and took them back to the lab for processing and sampling.

---

* Confession: Brown-headed cowbirds are my favorite bird. I've never had any maternal instincts either.

As soon as I opened the first cowbird bag, a delightful scent wafted by and I wondered which of my labmates had brought cookies. Then I realized that the scent was coming from the bag. That seemed odd. I sniffed the bird in my hand to confirm: yes, brown-headed cowbirds smelled like freshly baked sugar cookies! I ran around the lab shoving the poor cowbird in everyone's faces, demanding that they smell it. (Fun fact: People don't like it when you thrust a largish songbird in their face, particularly one with a rather strong and pointy beak.) All of the cowbirds I caught smelled this good.* My idea that perhaps cowbirds had no scent was clearly wrong.

So why don't birds abandon their nests when cowbirds lay eggs in them, or even eject the cowbird eggs? In my experiment, not a single bird abandoned its nest after smelling an unfamiliar bird's preen oil, even if it was from a different species. Even the startled female who picked up and took away nesting material came back after less than an hour. Furthermore, the reduction in time spent on the nest only applied to the very first time the females came back to the nest after I had applied any type of unfamiliar oil. When they returned for the next incubation bout, the difference disappeared. They went right back to sitting on their nests for their typical length of time before heading out for another meal.

I believe the answer is simple: odor fades. Volatile compounds are very small airborne molecules that evaporate quickly. After a short period of time, the odors from the visiting cowbird (or from experimentally applied preen oil) dissipate, and soon the junco will only detect her normal nest smell. It's also possible that the juncos just acclimate to the scent, just like humans with pets who don't notice the animals' odor in their house—even though

---

* In the interest of transparency: While I was editing this chapter, fellow preen-oil researcher Leanne Grieves tweeted at me, "IDK @juncostink, I've been smelling all the cowbirds & I'm not getting any cookie or baking vibes!!" I suspect she's sniffing them too early in the season. Maybe they will have a stronger odor in May or June.

visitors do. (I'm very familiar with this problem. I have four cats, two dogs, and a parrot.) An incubating female spends the majority of her time sitting in her nest, so any changes made to the smell would likely soon be overwhelmed by her own, more recently applied, odorous compounds.

BIRDS CAN SMELL, BUT DO THEY CARE?

At this point, I had discovered that birds can indeed smell odors produced by other birds. But it appeared that these scents don't last very long, and at least in the context of incubation, they don't have a lasting effect on behavior. Unlike the various excretions and secretions used by mammals for scent-marking, which are meant to be detected after the mammal is long gone (think dogs peeing on lamp posts*), the odors in preen oil are short-lived. I suspected that they were probably only important when the birds were near each other.

So, what do birds use these scents for, if anything? We know that in mammals odors are important for recognizing relatives, even if you haven't met them before, so that you can avoid breeding with them. Mammalian odors also send other kinds of information, like how healthy or aggressive the animal is. This kind of information can be important when choosing a mate, and it can help you to choose a high-quality, unrelated individual to mate with.

If birds use scents in the same way that mammals do, then the scents would have to contain individual information. Males should smell different from females, healthy birds should smell different from sick birds, and closely related birds should smell more alike than unrelated birds. The next stage of my research was going to be devoted to decoding the information in bird odor. But first, I was going to have to learn a lot more about chemistry.

---

* . . . and fire hydrants, and trees, and bushes, and fences, and interestingly shaped piles of snow, and . . .

CHAPTER 3

....

DECIPHERING THE
SECRETS OF SMELLS

Can't hear no buzzers and bells
Don't see no lights a-flashin'
Plays by sense of smell
— "Pinball Wizard," written by Pete Townshend,
performed by The Who

MEASURING SMELL

Studying smell comes with special challenges. We detect odors when odiferous airborne compounds bind to the olfactory receptors in our nasal lining. However, since olfactory receptors are controlled by hundreds of genes that each have thousands of variants, they are not exactly the same in each person. Each individual's unique combination of gene variants affects how the receptors are activated and how that person experiences specific odors. Even within a single person's lifetime, perception of odors can change, because hormone levels, age, and diseases can all affect the functioning of these receptors.

What happens in the brain further complicates the picture. When olfactory receptors are activated, they send information to the brain's olfactory bulb, passing through connections to the amygdala and hippocampus, triggering emotions and memories. Much more so than sights or sounds, smells are intimately connected to our feelings and recollections, affecting each person's experience of an odor.

People tend to be very bad at describing smells, even when it's a smell they recognize.* Our perceptions of odors can also be affected by the labels that are assigned to them, influencing whether we find them pleasant or not. For example, when exposed to the odor of a blend of isovaleric and butyric acids, study volunteers who were told the scent was parmesan cheese reacted very differently than those who were told it was vomit. Personally, I am extremely susceptible to suggestion when it comes to taste and smell, as I have noticed when tasting wine. At first, I just think "Hey, this wine smells and tastes nice," but as soon as somebody says "I taste notes of cherries" or "cocoa" or "petroleum," well, that's all I can taste. Professional sommeliers and perfumers go through intensive training to be able to identify and distinguish among individual odors.

Fortunately, we don't have to rely on verbal descriptions of smells in order to measure them. Chemists have developed highly sensitive methods to detect, measure, and categorize the molecules that activate olfactory receptors, allowing us to describe them in a standardized way. At the Indiana University Institute for Pheromone Research, Milos Novotny and Helena Soini rely on gas chromatography–mass spectrometry (GC-MS). There are several related techniques that are commonly used, but they all operate on the same general principle. These methods all separate different compounds in a gas on the basis of mass and electric charge, which is enough to allow us to identify the compounds.

---

* However, this problem may be cultural rather than universal. The Jahai people of the Malay Peninsula have a rich vocabulary for describing smells, and they name odors with ease.

GC-MS also measures how much of each compound is in the sample, so that we can describe the overall makeup of any odor mixture chemically.

The compounds identified in avian preen oil are mostly small- to medium-chain fatty alcohols and medium- to long-chain fatty acids, made up of different configurations of carbon, hydrogen, and oxygen atoms. At room temperature, the smaller compounds are liquid while the larger ones are waxy solids. The smallest compounds have a higher volatility, meaning that they have a higher tendency to vaporize at room temperature, and thus we refer to them as volatile compounds. The larger compounds are semivolatile and evaporate more slowly. Some of the compounds found in preen oil are very common in nature, and many of them are synthesized for commercial uses. For example, palmitic acid, an ingredient in the preen oil of many bird species, is the most common saturated fatty acid found throughout the bodies of both plants and animals. Palmitic acid is a major component of dairy products as well as plant-based oils, such as coconut oil and soybean oil. It's used in the production of soap and cosmetics, and also as a food additive.

In contrast to bird odors, most mammalian scent-marking involves heavy compounds. These compounds remain long after the animal that left them is gone, slowly releasing odors as the compounds break down so that other animals can still find the scent and get information about the signaler. Dogs aren't the only scent-markers; cats, for example, rub their cheek glands against the furniture, while hyenas rub their anal scent pouches against grass in their territory. The odors in these sticky scent-marks are often protein-based, like the major urinary proteins, or MUPs, in the urine of rodents and some other mammals.

Unlike the protein-based scents in mammalian scent-marks, the volatile compounds in preen oil are not direct products of the animal's genes. Protein-coding genes like MUPs can be identified

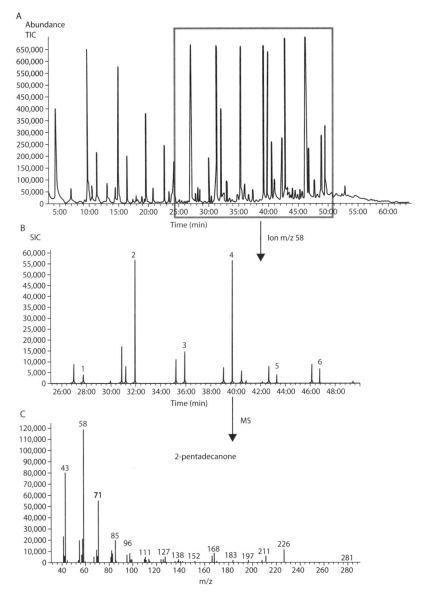

A gas chromatography–mass spectrometry (GC-MS) chromatogram of volatile compounds in a male junco's preen oil. Each peak represents a chemical compound; the height of each peak indicates how much of that compound was detected. A, The total ion chromatogram (TIC) of all compounds detected. B, A corresponding post-run single ion current (SIC) profile of methyl ketones with a mass-to-charge ratio (m/z) of 58, from the time range 25-50 min (1 = 2-dodecanone, 2 = 2-tridecanone, 3 = 2-tetradecanone, 4 = 2-pentadecanone, 5 = 2-hexadecanone, 6 = 2-heptadecanone). C, The mass spectrum of 2-pentadecanone from the SIC profile. *Adapted from Whittaker et al. 2010, Behavioral Ecology 21:608–614*

in the genome, and it is possible to test hypotheses about how these proteins have evolved over time. Odors that are directly linked to an animal's genome can communicate information about that individual's identity and quality. But when I first began trying to understand preen oil odors, no one knew how they were made, or whether they would follow the same principles as protein-based odors. Preen oil fulfills very different functions than mouse urine. Is it still possible for preen oil odors to reflect information about the birds? This was the initial question I wanted to address.

WHO ARE YOU? SPECIES, SEX, AND INDIVIDUAL IDENTITY VIA SMELL

The first piece of information that might be useful to an animal during social communication is whether its correspondent is a member of the same species, known as a *conspecific* in scientific lingo. Conspecifics are potential mates, potential cooperative group members, and potential competitors. Members of other species (*heterospecifics*) are generally not useful as reproductive partners or group members, and they are more likely to be predators, parasites, or prey.

In all modes of communication, conspecifics are more like each other than they are like any other species. For example, we can tell that all northern cardinals are the same species because they're all about the same size (1.5-1.7 ounces), all have a distinctive crest and red feathers (visual characteristics), and they all sing the same song types (auditory characteristics). Individual cardinals can recognize each other based on variation within these characteristics, but to the untrained eye and ear of other species (including humans), they all appear or sound the same. Chemical characteristics work the same way. Preen oil volatile compounds have now been described in many different species of birds. While there is quite a bit of overlap in the compounds found across species, each species has their own specific blend.

Just as with visual and auditory characteristics, these chemical characteristics vary within species. For example, all juncos have the same set of volatile compounds in their scent. However, the quantities of the compounds produced by each junco are different, so that each individual bird has its own unique version of the scent. This variation can include differences in how strong the total smell is, as well as proportional differences in particular compounds relative to the whole blend. Taste operates in a similar fashion, with the flavors of different foods and spices coming together to create an overall flavor profile in a dish. For example, chicken curry and shrimp scampi both have garlic, but in very different amounts relative to the other flavors—in the curry, garlic is one of many spices, but in the scampi, it is the primary note. In junco volatile profiles, we have not found any sex- or individual-specific compounds, but different blends are characteristic of males and females. During the breeding season, male juncos produce higher proportions of methyl ketones, while female odor blends are dominated by linear alcohols and carboxylic acids. Both are identifiably junco scents, and to humans, they seem to smell the same. But different aspects of the blend are accentuated in each sex, which may be how juncos—who are likely better attuned to subtle differences in the scent of their conspecifics than we are—can tell the difference between the scents of males and females.

In most of the bird species studied so far, sex differences in odor only seem to appear seasonally, when they are relevant for reproductive behavior. Like most birds living in temperate zones, juncos are seasonal breeders. In the winter, when they are concerned with surviving rather than reproducing, large numbers of both males and females live in flocks and forage for food together. There are no measurable differences between male and female junco scents in the winter.

But in the early spring, significant scent differences emerge between the sexes. Increasing levels of steroid sex hormones, such as testosterone and estradiol, power the reproductive process, initiating the production of sperm in males and eggs in females. These hormonal changes are reflected in the birds' behavior: males start singing, females start building nests, and the same birds that were content to live in large cooperative flocks throughout the winter start getting feisty with each other. At this time, the birds produce much higher concentrations of preen oil volatile compounds. While both sexes produce greater amounts of all the scent compounds overall, they also begin showing sex differences in which compounds are emphasized, likely driven by hormone production. Notably, birds may also have seasonal differences in their ability to detect variations in odors. A study of European starlings found that the birds had increased odor sensitivity to the scent of aromatic herbs—used in courtship and nest-building—during the breeding season, but they could not detect the same odor during the nonbreeding season.

In the spring of 2008, shortly before conducting my experiment on nesting female juncos, I took repeated samples from all of the juncos we could catch throughout the early breeding season at Mountain Lake. We used the samples to measure when and how quickly these seasonal changes in odor production occurred. In males, production of all compounds steadily increased in intensity over the first month of the breeding season, beginning around mid-April. These increases tracked the steady rise in male testosterone levels. In contrast, female junco scent suddenly spiked in concentration right around the same time that egg-laying began—for most females, that was during the third week of my study. Egg-laying was also when testosterone was highest in females (see figures on pp. 60–61). This link between scent and hormones means that each junco's smell is helpful to neighbors in determining not only whether an animal is an appropriate mate but also whether that potential mate is ready to breed.

Several years later, I learned that the sex differences in junco scent don't last for the entire breeding season. When we measured scents of parents who had just successfully raised a nest full of chicks, there were no differences between the paired males and females. At first, I was surprised and confused by the lack of a sex difference, which by this time I had taken for granted to be present throughout the breeding season. Had I made an error in my analysis? But then I thought about behavioral changes throughout reproductive cycles and realized that odor matched the overall pattern. Once the birds have successfully mated, they shift their energy and behaviors toward parental care. Sex hormone levels are lowered as both males and females care for offspring rather than seek mates. If the sex differences we observed in odor during the early breeding season are involved in stimulating partners to mate, those differences should go away once mating has been achieved, especially if the odors are costly in any way.

Age, another factor that can be important in choosing a mate, affects preen oil scent, too. In a study of gray catbirds, Dr. Rebecca Whelan of Oberlin University and her student Clara Shaw found that juvenile (adult sized, but not yet able to breed) birds produced a different odor profile than older adults, with more pentanoic acid, hexanoic acid, and heptanoic acid. I haven't measured older juvenile junco odors yet, but I did find that just-fledged nestling juncos had different odor composition than their parents, producing less 1-undecanol and more 1-pentadecanol and 1-hexadecanol than the adults.

Finally, as we know, just as individual birds have distinctive visual markings or specific variations in song, they also have unique scents. Researchers have even trained mice to differentiate the individual scents of birds. Behavioral studies have determined that birds can recognize the scents of known individuals. The differences in bird odors may not be detectable to us, but if the birds can

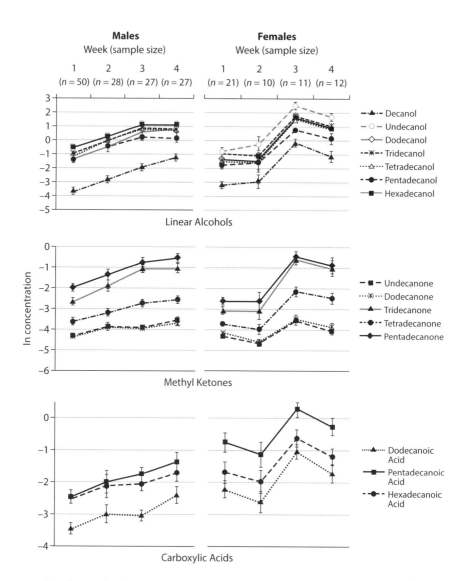

Abundance of different classes of volatile compounds in male and female junco preen oil during four weeks in early breeding season; n indicates the number of birds sampled at each timepoint. *Adapted from Whittaker et al.* 2011, Journal of Chemical Ecology 37:1349–1357

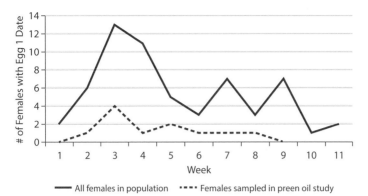

All females in population   ●●●● Females sampled in preen oil study

Frequency of females in the population laying eggs during each week of the breeding season. Note that the peak in egg-laying occurs in week three, the same week that female preen oil volatile compounds are at their highest concentration. *Adapted from Whittaker et al.* 2011, Journal of Chemical Ecology 37:1349–1357

recognize them, then there is the potential for these smells to be important in social behavior.

HOW ARE YOU? HEALTH AND HORMONES

In addition to communicating *who* an animal is, scents can also convey information about *how* the animal is, in terms of current condition and overall quality. An animal's current condition can be affected by disease, nutrition, and its own hormonal levels. "Quality" is usually defined as characteristics of an individual that correlate with reproductive fitness, and it often refers to static traits that don't change throughout an animal's lifetime, like its *genotype*. However, "quality" can also be used to refer to condition-dependent traits, like colorful plumage that correlates with how healthy or dominant a bird is. All of these pieces of information can be very useful to a bird when deciding whether or how to interact with another individual.

Infectious diseases can affect an animal's scent, providing crucial information about infection status to members of the same species. We know that just a few short hours after a human's immune response has been activated, other people can detect the change in that person's smell. Mice and rats can detect various viral, bacterial, and parasitic infections through olfaction, and they actively avoid the scent of infected conspecifics. As humans have discovered through long and sometimes painful experience, staying away from sick individuals is a good strategy to avoid getting sick yourself, so the ability to detect and respond to these odors is clearly advantageous.

A change in smell caused by an infectious disease could also be beneficial to the pathogen itself, particularly for a pathogen that relies on another animal, called a *vector*, to transmit it from one host to another. Malaria is one such disease, transmitted to vertebrate hosts by mosquitoes. Mammalian hosts infected with malaria produce odors that attract mosquitoes, increasing the pathogen's chance of being picked up and transmitted to another host. In humans, malarial infection changes skin odor, due to increased production of 2- and 3-methylbutanal, 3-hydroxy-2-butanone, and 6-methyl-5-hepten-2-one. Mosquitoes are more attracted to the skin odor of malaria-infected humans than to those who are uninfected. Similarly, malaria-infected mice produce overall greater concentrations of volatile compounds, making them more attractive to mosquitoes.

Malaria is caused by protozoan parasites from the genus *Plasmodium*, which have a complex life cycle involving different phases in their vectors and their vertebrate hosts. When a mosquito bites and takes a blood meal from an infected animal, the mosquito takes the parasites into its own body. The parasites mate and reproduce in the mosquito's gut but do not cause the mosquito any harm. After a period of about two weeks, the young *Plasmodium* parasites—called sporozoites at this stage—move from the mosquito's gut to its salivary glands. The next time the mosquito bites an animal, the sporozoites are injected into the animal's bloodstream. The parasites then migrate to the animal's liver for the next stage of growth. Once they reach maturity, they re-enter the bloodstream and multiply inside red blood cells until the cells burst, causing fever and anemia.* Malaria can be fatal, especially when caused by the species P. *falciparum*. There are about two hundred million cases of malaria in humans every year, mostly in tropical countries, causing about half a million deaths annually.

Avian malaria is caused by different species of Plasmodium that are adapted to birds, especially P. *relictum*, and it affects most species of passerine birds all over the world. It does not usually kill the birds outright, but it can affect their development, health, attractiveness to potential mates, and ability to produce eggs, as well as shorten their lifespan. Leanne Grieves, during her doctoral work at Western University in Ontario, Canada, was the first person to test whether malarial infection affected preen oil odor in birds. Grieves experimentally infected song sparrows with malarial parasites taken from the blood of already infected birds. She found that the overall odor profiles of infected birds were different from those of healthy birds. Although her methods could not identify the specific compounds that increased or decreased, her work was the first to show that parasitic infection changes bird odor.

We do not yet know for sure whether birds can detect the scent of infection in conspecifics. Grieves, using preen oil from infected and uninfected birds in her malaria study, tested whether captive song sparrows would avoid the scent of infected birds, whether they themselves were infected or not. She found that uninfected birds, and female birds regardless of infection status, showed a slight preference for the odor of uninfected birds, but the difference was not statistically significant. Perhaps the odor change is detectable by mosquitoes, attracting them to infected birds and

---

* My husband would just like me to remind you that he got malaria, and that I didn't, while he was assisting me with my fieldwork in Indonesia.

resulting in a spread of the pathogen. A recent study by a group of Spanish researchers found that mosquitoes were more attracted to the whole-body odor of house sparrows that had been infected with malaria. However, the mosquitoes displayed no difference in attraction to the uropygial gland secretions of these same birds. In contrast to Grieves's work on song sparrows, the researchers did not find a chemical difference between the preen oil of infected versus uninfected house sparrows, suggesting that the mosquitoes might have been attending to a different aspect of the birds' body odor.

Clues about the overall quality of an animal's immune system can also be present in odor. In particular, the MHC—the complex of immune genes that led to my interest in odor in the first place—is a key candidate for understanding how scent might be important in choosing a mate. Although much more remains to be discussed about MHC, for now, the important thing to know is that each individual has multiple copies of MHC genes, and that it's best if the copies aren't all identical. Each gene codes for a protein that can identify an invading pathogen and target it for the immune system. Slightly different genes produce slightly different proteins, which in turn can target different pathogens. Thus, if you have a "diverse" MHC genotype, you likely have a stronger immune system than someone who just has several copies of the same MHC gene. Following this logic, it's thought to be adaptive to seek out mates with diverse MHC (they are more resistant to disease) or to optimize offspring genotypes by preferring mates with different genotypes than one's own.

The relationship between odor and MHC is well documented in mammals and fish, but researchers only started studying it in birds within the last few years—probably because, like my colleague in the cafeteria, people operating under the assumption birds can't smell didn't think such study worthwhile. French researcher Dr. Sarah Leclaire found a correlation between MHC genetic distance and preen oil chemical distance when comparing pairs of black-legged kittiwakes, a type of Arctic gull. In other words, the more dissimilar two individuals were at their MHC genes, the more unalike they smelled. In his graduate work at Western University, Joel Slade, who later came to work with me as a postdoc at Michigan State, also found this correlation in song sparrows. To test whether the birds could detect these differences, his labmate, Leanne Grieves, who had done such pioneering work on avian malaria, gave captive song sparrows the choice to spend time with scent from an MHC-similar or an MHC-dissimilar bird of the opposite sex. She found that both males and females preferred the scent of MHC-dissimilar birds. Leclaire also tested odor preferences of another seabird, the blue petrel, but her results were not quite as expected: although 100% of the males tested preferred the scent of dissimilar females, 75% of the females preferred the scent of *similar* males. Clearly, we still have a lot to learn about MHC, odor, and mate choice in birds, but the studies so far show that information about immune-related genes does reside in a bird's scent.

As we noticed when considering the differences between males and females, hormones affect preen oil scent. Increases in steroid hormones like testosterone stimulate increases in volatile compounds, making scent a good measure of an animal's hormonal state and seasonal fluctuations in breeding condition. We know that many female mammals have a different scent when they are ovulating, and that males find the scent of ovulating females attractive. The same mechanism may be happening in birds.

Hormones are related to more than just breeding, of course. In general, male birds with higher levels of testosterone, which can sometimes be broadcast loudly through visual signals, are more territorial and aggressive than males with lower levels. There are many examples of species where males have physical characteristics that advertise their higher levels of testosterone: the black "bib" on a male house sparrow's chest is larger in males with higher testosterone, as are the white patches on the outer tail feathers (tail white) of dark-eyed juncos. These clear signs are ways to inform potential competitors that an aggressor is likely to lose, should they choose to attack, which lowers the risk of actual fighting. You might think that this would be an easy thing to fake, but as it turns out, these signals are kept honest because they are linked to successful survival and reproduction.

The vervet monkeys of eastern Africa are a particularly vivid example of honest signaling. The males have bright blue scrota, and the more dominant the male, the darker the scrotum. When dark and pale males are put together in pairs, the males with pale scrota avoid fighting by behaving submissively. But when two males with dark scrota are put together, they are much more likely to fight. When primatologist Dr. Melissa Gerald painted the scrota of pale males a darker blue and put them together with dark males, the painted males were attacked as if they were actually dark males. This goes to show that faking your dominance status can be very risky.*

Given the correlations we discovered between testosterone levels and preen oil volatile compounds in male juncos, and the relationships between testosterone and behavior, I began to wonder whether odor could signal information about a bird's behavior. Perhaps odor, like visual characteristics, could warn a competitor of how aggressively a bird would likely respond to them. Or perhaps a potential mate could assess, via scent, how well the bird would defend its territory. Can birds sense danger through smell?

* Interestingly, blue scrotal coloration in vervets is not related to testosterone levels. Instead, it is correlated with serotonin metabolites. The less dominant, pale males had lower serotonergic function—probably because they are more stressed than their dominant counterparts.

THE BEAR IN THE AVIARY

Toward the end of my years in the Ketterson lab, Dr. Kim Rosvall joined us as a postdoc. From the beginning, we got along very well, working together and meeting up for Friday "postdoc lunch." But the event that really created a lasting bond was the incident we still refer to as "The Bear."

In the spring of 2009, Ellen Ketterson had decided that it was time to replace many of the captive birds in the colony at the Kent Farm aviary at Indiana University. The birds that had been held for a few years were released at Mountain Lake, where they were originally captured. A small group of us volunteered to go to Mountain Lake later that year to capture and bring back seventy-five new juncos for the colony.

The Mountain Lake Biological Station is mostly closed in the fall, and that October the five of us were the only ones there, apart from the caretaker. Kim and I were joined by Christy Bergeon Burns and Mark Peterson, both graduate students in the Ketterson lab, and Ryan Kiley, a staff member who had overseen the Kent Farm aviary at IU for several years. It was great fun having the station to ourselves for a week or so. Every day we headed out to the country roads surrounding the station, where we set up mist nets at intervals and baited them with seed, trying to attract the winter junco population. We each took turns cooking dinner, and evenings were spent together around the large table in the communal kitchen, playing cards, drinking immoderately, and getting to know each other really, really well.

Catching seventy-five juncos in one week is ambitious, and some labmates back in Indiana were skeptical that we would be successful. We split up into two groups each day and got competitive about who could catch more. The nets were spaced out along back roads, and we would leapfrog along the nets, one group checking one net while the next group drove on to the next, and so on. The nets were often frozen in the cold mornings, and we would have to blow on them and rub them with our hands to melt the ice before we could open them for catching that day. On the last day, Ryan and I drove slowly past Kim and Christy, the opening bars of "The Final Countdown" blaring as we cheered each other on to meet our quota.

But not everything had gone happily up to that point in our junco-catching adventure. On our first day, we had caught two juncos in the morning, and we put them in the outdoor aviary that was in the woods a few hundred yards behind the laboratory building. We caught several more in the evening, and we were feeling pretty proud of ourselves. All of the birds had to be processed before they could be put in the aviary for the evening—we banded them, took body measurements, and took blood samples. It was starting to get dark, and it was Mark's turn to cook dinner, so he headed down to the kitchen. Ryan and I volunteered to finish processing the birds while Kim went for a run.

As we were finishing up and preparing to take the birds to the outdoor aviary, Kim suddenly appeared at the lab doorway, out of breath. "Bear," she gasped. "BEAR."

"What? Are you OK?" There were black bears in the mountains, but they were generally not a threat to us. However, they often raided the garbage—which is why there was an electric fence around the station dumpster—and could be frightening if you encountered one in the dark.

"Bear," she huffed insistently. Then, finally, she managed to shout, "Bear in the aviary!"

Apparently, when Kim had returned from her run, she guessed that it was about the time that we would be taking the birds out to the aviary, and so instead of going back up to the lab, she headed straight for the aviary to see if we needed help. It was getting quite dark in the woods, but she had a headlamp. As she approached the aviary, it was quiet, but she thought she saw someone in there.

"Ryan? Danielle?" she called. No response.

Kim Rosvall and the bear-damaged aviary.
Photo by Mark Peterson

She got closer. "Danielle?" And then suddenly there were two yellow eyes reflecting her headlamp light back at her.

Kim knew you weren't supposed to run when you encountered a bear, but run she did, all the way back to the lab. We were all upset by her news and needed to know how bad the situation was at the aviary. We went down to the kitchen to get Mark and Christy, and we made a pact that we would stick together and protect each other. We armed ourselves with flashlights and deliberately made our way to the aviary.

The bear had vanished, but not without first doing considerable damage. A whole panel of hardware cloth had been ripped off the side, and all of the containers of birdseed had been emptied and flipped over. The bear had obviously been looking for an easy meal before being scared off by Kim.

One of the two juncos we had captured earlier that day was still in the aviary, sitting immobile on a perch. (Songbirds don't have very good night vision and usually won't fly in the dark.) The second one had escaped but we guessed it hadn't gone too far. Ryan reached out carefully and gently grabbed the remaining junco.

Everyone was safe, but the aviary was ruined. We couldn't put our juncos in there, and there were only a few cages in the lab for temporary holding. Not to be deterred, Mark got in his car and sped down the mountain to a pet store in Christiansburg, hoping to get there before they closed for the night. He bought every single bird cage they had in stock, including the ones that were on display, which he had to awkwardly disassemble so he could get them in the car. The five of us spent hours assembling those cages, still fueled by adrenaline. Once the birds were all safely caged, we stayed up too late and drank too much beer, talking about the bear and trying to figure out how to proceed.

The year 2009 had been tough for some of us. That January, my parents had both died unexpectedly on the same day, twelve hours apart, of different causes. Kim had gone through a bad breakup earlier in the year. This bear incident felt like another link in a long chain of upsetting events. Kim and I talked about patterns in life—there are ups and downs, sometimes a lot of downs, but in general life follows an upward trajectory. That night she drew a line graph demonstrating this idea on a paper towel; I kept it pinned to the bulletin board in my office for many years.

Things did indeed get better. We caught the escaped bird in one of our mist nets a couple of days later—we knew it was him from the US Fish & Wildlife Service band we had tagged him with. And over the rest of the week, we continued to catch birds and house them indoors in the pet-store cages. In the end, we caught seventy-two juncos. Ryan and I drove the birds back to Indiana in my elderly Subaru. We packed the juncos up in brown paper grocery bags stapled at the top, three birds in each bag with orange slices and seeds in the bottom. It was an eight-hour drive from Mountain Lake back to Bloomington, and we had plenty of time to observe how the juncos reacted to music on the car stereo. The Violent Femmes led to junco squabbles in the bags, but they seemed to love Cyndi Lauper.

A few years later, a similar ursine invader situation arose. In the summer of 2013, I had caught eighteen juncos on which I was running behavioral experiments. But one day, another bear (or perhaps the same one) tore open the side of the aviary to get to the food, and half of my juncos escaped. The aviary is now surrounded by an electric fence to keep the bears out, and we have successfully run many more projects at Mountain Lake. One of these was a collaboration between Kim and me that grew out of a discussion during one of our evenings around the research station kitchen table. That project focused on a different kind of danger.

ARE YOU THREATENING ME? SCENT AND AGGRESSION

Many of the initial avian-scent-preference behavioral studies—including my own—focused on determining whether birds could distinguish between males and females based on scent, and which of those scents they preferred. A standard prediction is that birds in breeding condition will be attracted to the scent of the other sex, since we expect them to be looking for breeding opportunities. Of course, animal behavior is never quite this simple.

Spanish researcher Dr. Luisa Amo and colleagues conducted an experiment to test whether male house finches preferred the scent of females or other males. They put the male subject into the center of a specially designed cage with two ends for the bird to choose from. In one of the darkened enclosures at either end of the cage, the researchers placed another, hidden, male bird, and then they placed a hidden female at the other end. Air flow pushed the scents from both ends toward the bird in the central chamber. The researchers recorded which side the male chose to enter.

To their surprise, only about half of the males chose the side of the cage near the female scent; the other half preferred to head for the side containing another male. If the male birds were more attracted to female scent, one might conclude that odor is important in finding a mate. On the other hand, if they were more attracted to male scent, perhaps the scent of another male incited territorial responses, with the male subject approaching the scent looking for a fight. However, different males made different choices. What did it mean?

The researchers re-evaluated their study. Perhaps there was something about the individual "scent donors" that affected the results. They measured the body condition,* immune function,† and feather coloration‡ of each. They found the choices of the test subjects depended not on sex but on the quality of the male scent donors. If the male choosers in the center of the cage had better body condition and immune response than the male scent donors at one end, then they chose to approach the male side of the cage. However, if the male scent donor at one end had better body condition than the chooser, then that test subject steered clear and instead headed for the end with the scent of the female. The researchers concluded that male house finches can use scent to evaluate potential rivals, and, based on that information, the birds decide whether they want to approach and fight the rival.

How could an animal's odor lead a competitor to decide whether or not to attack? We know that testosterone is often correlated with aggression, as its main effect on body tissues is to shift resources to greater activity and energy mobilization, leading to changes in many physical and behavioral characteristics. When a gland increases production of any hormone, it sends signals to other body organs to change what they are doing. For example,

---

* Body condition is determined by the ratio of mass to tarsus length. The heavier the bird relative to its skeletal size, the better its condition.

† A common way to measure immune response in birds is to inject phytohemag-glutinin (PHA) into the skin inside the wing. PHA is a lectin found in red kidney beans that induces swelling. Other than some mild discomfort, PHA has no adverse effects on the bird . . . unlike the effects on the human digestive system caused by eating undercooked kidney beans. (Ask me how I know.)

‡ House finches have red chest feathers that reflect the carotenoids—red and orange pigments that provide antioxidant benefits—in the plant foods they eat.

when the adrenal glands secrete adrenaline, they are telling the body to get ready to either fight or flee. Multiple systems respond: the animal's heart rate and blood pressure increase, the pupils dilate, and the air passages expand. In order to receive these signals, the different body tissues all have specific receptors for certain hormones. When a male's testes increase production of testosterone (one of several androgen hormones in the body), the testosterone is released in higher concentrations into the blood stream, and the blood stream carries the testosterone to all of the tissues in the body. Tissues lacking androgen receptors would not be affected at all. The more receptors a tissue has, the more sensitive it is, meaning it can respond to hormone-level changes more quickly or strongly. In the tissues that do have androgen receptors, testosterone binds to the cell receptors, triggering changes in cell function by turning certain genes on or off.

We often think of violence and aggression as male traits and consider females to be nurturing and demure, but that's really just social conditioning on our part. Although males typically have higher levels of testosterone than females, females do not lack this hormone. Animal behavioral scientists have long assumed that aggressiveness was costly to females because it would take energy away from reproduction and cause unnecessary exposure to risk. Instead, males were thought to handle the territorial defense while females cared for the offspring.

However, that assumption doesn't always track. Some female animals need to be aggressive in order to survive and reproduce. For example, tree swallows build their nests in cavities, such as hollowed-out holes in trees, and also readily use nesting boxes built by people. They are obligate secondary cavity nesters—meaning that they can't dig holes but instead must reuse cavities that other animals have created in the past. Because females cannot hollow out their own nest sites, the number of available cavities limits the number of females that can build nests. No nest means no reproduction that year. Thus, females must be very aggressive in defending their nests from other female tree swallows.

Kim Rosvall, who is now an associate professor at Indiana University, is interested in how aggression and competition is regulated in female birds, despite lower levels of testosterone. To test whether aggression is costly to female tree swallows, Kim removed nest boxes from a tree swallow nesting site, which meant that the females would have to fight harder to commandeer the remaining boxes. She found that the more aggressive females were indeed more likely to snag a site. Interestingly, the aggression levels of their male mates had no effect on whether the pair was successful. Kim concluded that aggression could be adaptive for females, contrary to conventional wisdom.

High circulating levels of testosterone can be harmful to females, as studies on juncos have shown. Female juncos with experimentally increased testosterone were less likely to build nests and lay eggs, and those that did spent less time caring for their offspring, resulting in reduced survival. How can females balance the benefits of testosterone—such as aggressive behavior when you need it—with the costs?

Kim decided to look at this question in dark-eyed juncos, focusing on hormone receptors in areas of the brain related to social behavior and aggression. First, to measure aggression, Kim conducted a kind of behavioral trial called a *simulated territorial intrusion*, or STI* for short. The intrusion is simulated by researchers setting a caged bird in the middle of a wild bird's home range, alongside a speaker playing a recording of bird song. When the wild bird nearby hears the song, it will react as if another wild bird had shown up in its territory and attempted to claim it. This reaction

* Not to be confused with sexually transmitted infections.

usually includes attempts to intimidate the intruder by swooping over the cage repeatedly, singing, and generally spending time in close proximity to the cage, which is in itself threatening. The more time spent near the cage and the more swoops and songs performed, the more aggressive the response. Kim targeted birds of both sexes who were in the incubation phase of reproduction, when we would expect them to be the most aggressive and territorial. After the trials, she measured testosterone levels in the blood, as well as expression of receptors and enzymes in the brain. After sacrificing the birds quickly and humanely, she dissected the brains and extracted RNA from brain tissues. Using a technique called quantitative polymerase chain reaction (qPCR), she measured the quantity of androgen receptors, estrogen receptors, and the enzyme aromatase, which converts testosterone to estradiol.

Both sexes reacted aggressively toward the intruder, and there was no sex difference in their aggression levels. As expected, the higher a male's testosterone, the more aggressive he was toward an intruder. But this relationship did not exist in females. Kim found that in both males and females, the aggressive response correlated with their expression of androgen receptors, estrogen receptors, and aromatase in the ventromedial telencephalon, an area of the forebrain involved in competitive and parental behavior. Thus, the more sensitive their brains were to hormonal changes, the stronger their response to the hormones. By expressing more receptors in the brain, females could increase their sensitivity to testosterone without creating too much of it.

After bonding over the bear-in-the-aviary incident, Kim and I had toyed with the idea of collaborating on a project, but we hadn't found the right idea yet. Once I was making real progress on understanding what information was present in bird scents, Kim saw a great opportunity to team up. She asked me if I thought birds might also signal their aggression levels through odor. Based on scent alone, would potential rivals be able to decide whether it was a good idea to attack? Does testosterone affect preen oil odor? What about in females, who have lower levels of testosterone but are still aggressive? Perhaps odor differences relate to androgen receptor expression, which would affect males and females similarly.

We decided to compare odor, testosterone levels, and androgen receptors in the uropygial gland with aggressive responses in both male and female juncos, using the same STI behavioral trials and qPCR analyses that Kim had used before. The more aggressive males not only had higher testosterone levels but their uropygial glands had more copies of the androgen receptors, making them more sensitive to changes in testosterone. Male scent was strongly correlated with the bird's testosterone levels, how often he dive-bombed a caged male, and how many songs he performed. The more aggressive males also produced higher concentrations of several volatile compounds in their preen oil. We concluded that scent communicated aggressiveness in male juncos.

What about females? Females had significantly lower levels of testosterone than males, as expected. Neither their scent nor their overall aggression levels were correlated with testosterone levels. But females had just as many androgen receptors in their uropygial glands as males, suggesting that the gland should react to hormone signaling just as strongly as in males. And we found correlations between their scent and one measure of aggression: how much time they spent responding to an intruder. It appears that, just as in males, scent does indeed signal aggression in females.

Scent is clearly useful in understanding differences among individuals. In addition to information about an individual's identity and condition, odors can tell you something about how an individual might behave in response to a social situation. However, what about cases where one entire population is more aggressive or otherwise behaves unlike another? Do the animals in these populations smell different from each other? One special urban population of juncos provided me with the opportunity to explore just this question.

In San Diego County, California, a population of juncos has recently evolved a new way of life. Although it's common in many parts of North America to see juncos at bird feeders in the winter, during the summer breeding season they are rarely seen in urban or suburban environments. They prefer to breed in cool, wooded areas with lots of evergreen trees, especially in the mountains. Then, when the winter arrives, they migrate to milder climates. While some populations of juncos, like those found in Canada, are long-distance migrants who fly south for the winter, others that breed in mountains at southern latitudes simply move down the mountains to the lowlands.

The Laguna Mountain range in San Diego County is typical junco breeding habitat, with elevations of about six thousand feet and abundant evergreen trees. The University of California at San Diego campus, about an hour's drive across a desert, has long been in the junco winter range. Like many college campuses, UCSD is full of concrete, people, bicycles, garbage cans, and introduced plants like eucalyptus trees. It is also located on the coast and has mild weather year-round, often described as Mediterranean. One spring in the early 1980s, a few pairs of juncos didn't make the trip back to the mountains. Instead, they stayed for the summer and established breeding territories on campus.

Now, about 40 years later, there is a stable and successful population of resident juncos at UCSD, and others in various suburban areas in southern California. This change in habitat has brought new selection pressures to such populations: the warmer climate means that they can produce up to twice as many broods of nestlings each season, while the abundant resources have led to denser populations and smaller territories. In just a few decades (a very short time, evolutionarily speaking), the UCSD juncos have adapted to these new pressures with changes in their behavior, hormones, and even their body shape and plumage.

Many of these changes were first documented in the early 2000s by Pamela Yeh, then a doctoral student working with Dr. Trevor Price at UCSD. Later, Jonathan Atwell, as a graduate student in Dr. Ellen Ketterson's lab at Indiana University, set out to catalog as many differences between the UCSD population and the ancestral population in the mountains as possible, and to try to understand what was controlling these differences. Since the shifts in behavioral and physical traits all came about at the same time, it seemed likely that a single physiological change could be involved with them all. Atwell suspected that evolutionary shifts in the production of testosterone—because it is involved in so many reproductive behaviors and is known to be correlated with multiple physical features—could be driving this change.

The climate at UCSD allows for a longer breeding season, and the juncos there have shifted their hormonal levels accordingly. Atwell found that the UCSD males experience the seasonal rise in testosterone much earlier than their counterparts in the mountains, getting their bodies ready for reproduction as early as March, compared to mid-April in the Laguna Mountains. These testosterone levels stay elevated for much longer in the UCSD juncos (average 125 days, compared to only 75 days in the mountain population). However, even though the testosterone levels of the urban juncos are elevated for a longer time, at their highest levels, the mountain juncos produce significantly more testosterone than the urban birds.

UCSD males exhibit markedly different behavior than the mountain juncos, spending more time helping take care of their nestlings and less time mating with extra-pair females. This shift in behavior increases their chances of successfully raising their nestlings—as many as four broods per season, compared to only two in the mountains. With the increase in caring behavior came a decrease in territorial behavior. The urban juncos responded much less aggressively to intruders than the mountain juncos did. Considering the urban juncos live at much higher densities, with smaller territories much closer together than seen in the mountains, reduced aggression toward each other is likely an important factor in their success story. Less time fighting means more time for reproduction, and lower chances of injury and mortality. Because parental care, extra-pair mating, and aggression are all known to be correlated with testosterone levels in many animals, the shift in seasonal testosterone secretion patterns Atwell observed in the urban population at UCSD could account for all of the stunningly rapid behavioral changes in this population.

But the accelerated transformation the UCSD juncos were undergoing wasn't strictly behavioral. Their bodies even changed in this short time. They have shorter wings relative to body size, compared to their country cousins, a trait associated with a reduction in migration. Their plumage coloration also changed. The juncos on the West Coast of North America belong to the subspecies *Junco hyemalis thurberi*, often called Oregon juncos. While the "slate-colored" subspecies on the East Coast (*J. hyemalis hyemalis*) is primarily dark gray with a white belly and white outer tail feathers, the Oregon subspecies is much browner, with black feathers on their heads. Male UCSD juncos have less extensive head black and less tail white than their counterparts in the mountains. Both head black and tail white are thought to be attractive to females. We know from the Mountain Lake population of juncos in Virginia that tail white reflects testosterone production, again suggesting that a hormonal shift could be the single evolutionary change driving the profound and rapid differentiations between the urban juncos and their ancestors.

Urbanization also comes with its own challenges—namely, dealing with humans and their environments. To cope, many animals that live in urban areas have lower levels of corticosterone. Chronically high levels of stress-related hormones can have severe consequences for an animal's health—as many humans are all too aware—so keeping these levels low is an important adaptation for surviving in a stressful environment. Atwell found that this was true of the UCSD juncos: compared to the Laguna Mountain juncos, their levels of corticosterone did not increase as much in response to being handled by a researcher. They were also less easily spooked by the approach of a human, allowing a person to get as close as thirteen feet before flying away, compared to about forty-two feet in the mountains.

Finally, Atwell wanted to know whether all of these changes were truly due to genetic evolution, or whether they were developmental. In other words, did simply growing up in an urban environment affect the juncos' development, causing them to secrete lower levels of testosterone and corticosterone and thereby changing their bodies and behaviors as adults? Or did the environment select for different genes over several generations? To find out, he used a classic method called a *common garden experiment*. He captured very young juncos at both locations and brought them back to the aviary in Indiana to raise them in identical conditions—same kinds of enclosures, same food, same light schedules, everything. When raised in the same environment, did birds from the two different populations still exhibit the divergences that were witnessed in the wild?

In short, yes. The captive UCSD juncos were still smaller and had less head black and tail white than the captive mountain juncos. The captive juncos from UCSD still went into breeding condition earlier and for longer than those from the mountains. As in the wild, the captive urban juncos' peak testosterone levels were lower than those of the captive mountain juncos. They showed the same differences in stress levels that we observed between the

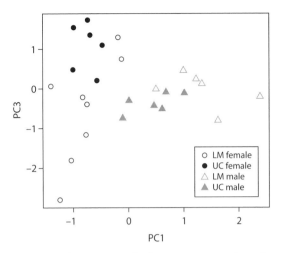

Plot showing principal component (PC) scores representing odor measures in juncos from Laguna Mountain (LM) and University of California San Diego (UC). Note that the two populations are different, and within each population, males and females are different. *Adapted from Whittaker et al. 2010, Behavioral Ecology 21:608–614*

wild populations. The UCSD juncos were also bolder and more exploratory when placed in an enclosure they had never seen before. You may not be shocked to hear that all this made me curious about whether juncos from the two populations smell different, too. I collaborated with Atwell to explore whether the juncos' odor reflected their genetic background, or whether it instead was more heavily affected by their living conditions. We found that, despite living in identical conditions, the birds from the two populations were distinct in their odors. They all had the same compounds present in their preen oil—not surprising, since they were all from the same species and even subspecies—but the relative amounts of certain odiferous compounds were different. When I plotted out measurements of each individual's preen oil composition, there were distinct clusters: the males all clustered together, as did all the females, but within each sex, there was a clear UCSD cluster and a Laguna Mountain cluster. We concluded that there must be a genetic component to the birds' odors, and that birds should be able to differentiate between members of their own population and birds from a different, closely related population.

The differences I observed between sexes, populations, and individuals were subtle but statistically significant. Could the juncos differentiate between these fine gradations in odor? Did they care about them? The only way to find out was to ask the juncos themselves.

HOW TO KNOW WHAT THE BIRD'S NOSE KNOWS

Since I began studying and thinking about bird scents, I've become much more attuned to the odors I encounter both in my work and in my life. I've noticed that different species of birds have distinct odors: brown-headed cowbirds smell like cookies, red-bellied woodpeckers smell awful, and dark-eyed juncos smell like leaves and dirt. But I can't detect the subtle differences between the scents of individual juncos. Instead, I have to use chemistry and statistics. However, just because we can detect differences between odors using highly sensitive machines like gas chromatographs doesn't necessarily mean that these odors matter to the animals.

Some researchers have trained mammals to discriminate between one bird's odor and another's, giving us the first indication that there are biologically detectable scent differences between individuals. In a study of red junglefowl (the ancestor of domesticated chickens), Swedish researcher Anna-Carin Karlsson and colleagues trained mice to differentiate among individual bird body odors using standard operant conditioning techniques. They trained the mice by presenting them with the odor of one bird and rewarding them whenever that bird was present. Then, the mice were presented with multiple bird scents; a "correct" response meant the mouse licked a water tube only if it was presented with the bird odor the mouse had been trained to recognize. All of the mice were able to make the correct response over 85% of the time, demonstrating that the birds had individual body odors that could be recognized by another animal.

Similarly, French researcher Aurélie Célérier found that mice could discriminate between the scents of individual blue petrels. In the habituation-discrimination method, animals are first presented with a reference odor long enough to get familiar with it (habituation). In the second phase, they are given two odors: the familiar reference odor and a different, unfamiliar odor. If they spend more time investigating the unfamiliar odor than would be expected by chance, we can conclude that they can tell the difference. In Célérier's trials, the mice discriminated between different adult petrels, even if they were the same sex, as well as individual nestlings and fledglings. Interestingly, she also found that related individuals had similar scents. Mice that had habituated to the scent of a nestling also treated the scent of that nestling's parent as a "familiar" odor, spending more time investigating the scent of an unrelated bird.

These studies show that the differences between individual bird odors can be detected by animal noses—at least, by mammalian noses. The real question, however, is can the birds themselves tell each other apart using smell? And if so, are these odors really used in bird communication? If we find that they do, we can further explore the possibility that evolutionary processes like natural and sexual selection have shaped the odors to be attractive or repellent. You can train some animals, such as chimpanzees, to answer questions by pushing a button or clicking a touchscreen, and as we've already seen, you can train laboratory mice with operant conditioning. But for most other animals, especially wild animals, sometimes our methods of inquiry have to be a little more creative.

Many of our assumptions about how animals use odor are based on the idea that some of these odors will be more attractive than others. So, rather than using habituation-discrimination techniques to test whether animals can differentiate between scents, we can go one step farther and test their preferences when given a two-way choice. A commonly used method is a Y-*maze*, so called because it is shaped like a Y, with a neutral area at the base of the Y, where the test animal is placed, and the two choices located at the end of the Y arms, one in each arm. A clear behavioral preference for one odor over another is an indication that the test animal can tell the difference between the two.

Francesco Bonadonna and Gabrielle Nevitt used a Y-maze to test the olfactory preferences of Antarctic prions, seabirds that have white eyebrows and blue feet and that eat primarily zooplankton, which they obtain by filtering water through their upper bills, almost in the way some whales use baleen. Prions live in burrows with their mate, in large, dense breeding colonies. As they often return to the burrow at night after foraging at sea, they must use their sense of smell to locate the correct burrow. When given the choice between the scent of their partner and the scent of another, random prion, seventeen out of twenty prions tested chose the scent of their partner. This paper, published in Science in 2004, was the first demonstration that birds could recognize the scent of an individual. These same methods have been used in many studies, including a few already mentioned in this chapter: Luisa Amo's study on whether male house finches preferred the scent of males or females and the studies of whether song sparrows and blue petrels prefer preen oil from MHC-similar or MHC-dissimilar birds.

I first used this protocol to test whether juncos could detect the same volatile differences that I had identified using chemical methods. Ryan Kiley, my fellow bear-in-the-aviary adventurer, built a songbird-sized Y-maze out of plexiglass, which would allow us to see inside. At the end of each Y arm, we put a cotton pad with a few drops of a mixture containing preen oil and acetone. The acetone dissolved the waxes in the preen oil and acted as a carrier

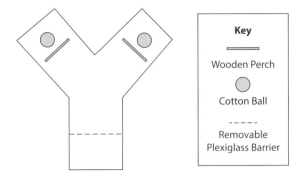

Diagram of plexiglass Y-maze used in junco odor preference trials. *Adapted from Whittaker et al. 2011*, Behavioral
Ecology 22:1256–1263

solution for the volatile compounds, just like the alcohols in perfumes that evaporate and leave only the heavier scent behind. Our
study subjects were captive birds from four junco populations:
the two southern California populations of Oregon juncos (one
urban, one mountain), the Carolina subspecies (J. h. *carolinensis*)
from Mountain Lake Biological Station in Virginia, and the white-
winged juncos (J. h. *aikeni*) that Christy Bergeon Burns (who had
also been with us for the bear-in-the-aviary incident) had driven
back from South Dakota.

For the first two tests, the subjects were only given the choice
between two birds from their own population. In the first trial,
I gave juncos a choice between preen oil from a member of the
same sex and opposite sex. Since I thought that odor was probably
important in mate choice, I expected that males would be more
attracted to the scent of females and vice versa, especially since
all of the birds were in breeding condition. To my surprise, males
strongly preferred the scent of other males. The results for females
were a little less clear, but they tended to spend more time with
the odor of males than females.

In another set of tests, I wanted to know whether males who were
visually attractive—in this case, males with more tail white—also

had more attractive scents. If attractive traits are honest indicators that the animal is healthy or has good genes, then we would
expect an animal to be attractive or unattractive in multiple sensory modalities. I expected that females would prefer the scent of
males with more attractive plumage compared to less attractive
males. But I was wrong again. Females seemed to prefer the scent
of males with less tail white, although the differences between the
two choices were not statistically significant.

Finally, I tested for population preferences. I gave female subjects the scents of two different males, and male subjects got to
choose between the scents of two different females. I predicted
that if scent was important in mate choice, then the birds would
prefer the scent of potential partners from their own population.
Although the many subspecies of juncos are capable of interbreed-
ing, they rarely do. I presented the test subjects with odors from
a bird from their own population and a bird from a different population. I tested our four populations in two pairs: the two closely
related populations from San Diego County were one pair, and the
distantly related Virginia and South Dakota populations formed
the second pair. Here, the results were even more confusing.
Females from the urban, UCSD population preferred the scent
of males from their own population—but females from Laguna
Mountain also preferred the scent of UCSD males. Females from
the Virginia population preferred the scent of Virginia males,
and so did the South Dakota females. (Males from all of the populations showed no preferences at all, which was less surprising,
since males are generally expected to be less "choosy" about who
they mate with.) What was going on here? I was frustrated by
these puzzling results but also intrigued—I suspected I was missing something interesting.

In animal behavior, when your predictions are all so spectac-
ularly wrong, it's time to take a long, hard look at your study and
ask yourself what you are really testing. I designed the study with

questions about mate choice in mind, but a bird in a plexiglass maze with some scented cotton balls is not choosing a mate. The bird is probably a bit stressed and confused and not interested in sex at all. Perhaps the birds' choices were related to a different instinct, such as territoriality or fear.

I went back and analyzed the characteristics of the males whose scents the females preferred, and I compared them to the rejected scents. Specifically, I compiled data on the males' tail white (attractiveness), wing length (often used as a proxy for body size), and ratio of body mass to tarsus length (a measure of body condition—basically, how heavy the bird is relative to its body frame). A pattern emerged: whether choosing from within their own population or between their own and a different population, females preferred to be near the scent of smaller males, with shorter wings and lower mass to tarsus ratios. (There was no relationship at all with tail white.) UCSD juncos are slightly smaller than Laguna Mountain juncos, and females from both California populations preferred the scent of the UCSD males. White-winged juncos from South Dakota are one of the largest subspecies of juncos, with males in our study having wings of about 3.5 inches compared to 3.2 inches in the Virginia juncos. Both Virginia and South Dakota females strongly preferred to be near the scent of males from Virginia and avoided the South Dakota male scents.

Was this a coincidence, or could the birds differentiate between the scents of larger and smaller males? Were there detectable odor differences related to size? I reanalyzed my GC-MS dataset from a different study and found a significant relationship between size and preen oil volatile compound profiles. Males with longer wings had more of the compounds that were typically higher in males—they smelled more "male-like." So yes, the chemical information was definitely present.

I considered a few plausible explanations for these results. In the wild, male junco wing length is positively correlated with reproductive success, meaning that males with larger wings have more offspring. In this study, females preferred the scent of males with smaller wings, but they might be less likely to mate with those males if they met them. The explanation I liked most at the time was that smaller, less attractive males "overcompensate" by producing an attractive scent. In our press release about the paper, I called it the "avian Axe effect," attempting a cheesy reference to a particularly odiferous men's body spray that was heavily advertised at the time.

But I'm no longer certain that's actually what was going on. The choice of a scented cotton ball in a Y-maze may not have anything to do with sexual attraction. The males in the study overwhelmingly preferred the scent of other males over the scent of females, which seems like a territorial response. Females may also be reacting territorially. After all, both male and female juncos in the wild respond aggressively to intruders in their territory. Perhaps, like the males in Amo's study of house finches, the females in my study decided they would choose to attack the smaller males, rather than mate with them. Or, rather than being attracted to the scent of the smaller males, they may have just been avoiding the threat posed by the scent of the larger males.

As I had hoped, my experiment did demonstrate that juncos could tell the difference between the scents of different juncos. They could use the information coded in smell to make decisions about behavior. But despite my best intentions, the experiment did not reveal anything about how birds might use scent when evaluating potential mates. I needed to know more about the mate choices birds were making when not negotiating Y-mazes in labs, so I decided it was time to pack up my gear and head back to the field.

CHAPTER 4

WHAT DOES SEXY SMELL LIKE?

Which organic sense is the most ungrateful and also seems to be the most dispensable? The sense of smell. It does not pay to cultivate it or refine it at all in order to enjoy; for there are more disgusting objects than pleasant ones (especially in crowded places), and even when we come across something fragrant, the pleasure coming from the sense of smell is always fleeting and transient.
　　—IMMANUEL KANT, *Anthropology from*
　　　a Pragmatic Point of View

TIMING IS EVERYTHING

To ensure evolutionary success, the best time to mate is obviously when mating is most likely to lead to successful reproduction. Specifically, that is when the female's eggs are ready to be fertilized and the male has sufficient sperm available to fertilize them. Changes in sex hormone levels that initiate ovulation and spermatogenesis also influence an individual's smell. Scent is therefore a reliable way for males and females to detect each other's readiness and synchronize their reproductive timing.

Long-range attractants that advertise readiness are especially important in solitary species, since they may not encounter many suitable partners without them. Which sex emits these chemical attractants? It depends on the species, and on which sex is the limited resource. In mammals and birds, females are usually the limiting sex, since they produce a limited number of eggs, which can only be fertilized by one or a few males. The males, in contrast, can produce copious amounts of sperm and inseminate as many females as they can access. Females must invest more energy into producing eggs and raising offspring, while males are able to expend their energy searching for potential mates. For this reason, females are often the sex that produces long-range chemical signals. For example, in most species of insects, females secrete long-range attractant pheromones. In silkworm moths, males can detect a female's pheromone from several hundred yards away. Female domestic dogs are known to produce long-range attractant pheromones when they are "in heat" (a fact well-known to many dog owners), although scientists have not yet identified the specific chemicals that make up these pheromones.

But there are exceptions, like the burying beetle. These beetles depend on small animal carcasses (usually a bird or a rodent) for food to raise their offspring. The male burying beetle locates and guards a carcass until a female comes along. The female helps to prepare the carcass by removing any fur or feathers, rolling it, and smearing it with anal and oral secretions. The female then lays her eggs near the carcass, so that, upon emerging, the offspring can use this nutritious resource to grow. Because the males are limited in how many carcasses they can guard, the males are the limiting sex, so they are the ones to send out the long-range attractant signal. Similarly, in lampreys and gobies, males find and protect nesting sites before sending out a chemical signal to attract females, who assess the site and decide if they want to lay eggs there.

Long-range attractants seem to be less important in social species, where the sexes encounter each other regularly. But they still need to get the timing right. Females of many species typically have a short window of only a few days in which their eggs can be fertilized. In many mammals, from hamsters to pandas, females increase scent-marking activity just before or right at the beginning of their reproductive period, known as estrus. Estrus occurs when the lining of the uterus has thickened in preparation for a fertilized egg. At this time, a female's progesterone and estrogen levels increase her behavioral receptivity to potential mates, and the female's chemical signals indicate to males that she is ready to mate.

The scent of a male can also induce hormonal and behavioral changes in female mammals. In pigs, livestock scientists have found that male saliva contains a pheromone that attracts females and stimulates mating behavior. Female mice investigate male scent-marks before mating. Many studies show that if you impair a female's ability to smell, her sexual behavior is reduced.

In birds, more preen oil volatile compounds are produced in the breeding season than in the nonbreeding season. The composition of preen oil switches from heavier, waxy compounds to lighter volatile compounds that give off more scent. Female juncos increase overall preen oil volatile compound production right around the time they begin laying eggs. Shortly after the egg-laying period, production drops sharply. Sex differences in junco scent are at their most distinct in the early breeding season, but they completely disappear by the time the parents are done raising their chicks. The timing of the increase in odor production, as well as the differences between males and females, suggests to me that females use these scents to attract males, or to advertise that they are ready to mate.

Male bird behavior also changes during the early breeding season. Males increase mate guarding behavior around the time of the female's ovulation. Such vigilance is necessary if the male wants to ensure that he raises only his own offspring, because neighboring males are also paying attention to the female's cues and looking for opportunities to mate with her. In mammals, we know that exposure to the scent of a female who is ready to mate increases testosterone levels in males. Increased testosterone leads to increased sperm production, as well as enhancing the male's own readiness to mate.

The first evidence that avian scents may affect reproduction was found by Belgian researchers Jacques Balthazart and Ernest Schoffeniels in a study of domesticated mallard ducks in the late 1970s. The researchers experimentally severed the olfactory nerve in ten young male mallards and compared their behavior to that of ten sham-operated males. Sham operations are common control conditions in which the animals undergo much of the same procedure, including anesthesia, but without the actual experimental operation, so that any side effects of having had surgery can be ruled out as contributing factors. Balthazart and Schoffeniels found that the males who lacked a sense of smell displayed normal behavior in most ways except for sexual behavior: they almost never attempted to mate with females.

Thirty years later, Japanese researcher Atsushi Hirao and colleagues set out to test whether female uropygial secretions affected male sexual behavior in chickens. Each male chicken was placed in a pen with four females in breeding condition, two of which had had their uropygial glands removed and two of which had not. Although the males courted all of the females equally, they were far more likely to attempt to mate with females with intact uropygial glands. It seemed that the females' secretions stimulated the males to mate. Was the effect caused by scents in the uropygial secretions?

The researchers repeated the experiment, but this time they bulbectomized half of the males: that is, using a suction pump,

A gray catbird biting my hand. The bird's right naris (plural nares), or nasal passage, is clearly visible on the upper beak.

they removed the olfactory bulb from the male chickens' brains through a small hole in the skull. Although the procedure sounds gruesome, the birds were observed carefully and found to behave normally, apart from being unable to smell. These twelve males, plus twelve control males, were each presented with two glandectomized and two intact females, just as in the previous experiment. Again, the males with a sense of smell strongly preferred to mate with the females that had uropygial glands. But the bulbectomized males were equally likely to attempt to mate with any of the females—the females' lack of a uropygial gland had no effect on the sexual behavior of the males without olfactory bulbs.

Finally, Balthazart, along with Mélanie Taziaux, conducted a study with Japanese quail to understand what changes happened in a male's brain when exposed to the scent of a female in breeding condition. Two groups of males were allowed to copulate with females. One group had their *nares* (nostrils) plugged, while the other group did not. Ninety minutes later, the researchers collected the males' brains, sliced them, and stained them using a technique called *immunohistochemistry*, which targets the protein product of an immediate early gene called c-fos. Immediate early genes are activated whenever a cell responds to a stimulus from outside the cell, and, as the name suggests, they are activated quickly and before other genes can respond. Researchers can measure this activation and determine which part of the brain responds to particular events. The products of these genes, which may include transcription factors, DNA-binding proteins, or other units important in building proteins, are then used for the cell's response to the stimulus. Balthazart and Taziaux found that copulation in both groups of Japanese quail activated two areas of the brain (the medial preoptic area in the hypothalamus and the bed nucleus of the medial stria terminalis in the limbic forebrain). However, males whose nostrils had been blocked had significantly lower expression of the fast-acting immediate early genes in these brain areas. Again, smell seems to be important in the biology of mating behavior in birds, just as in mammals.

Chemical signals can be used to attract potential mates and get their physiological states in sync to ensure successful fertilization, but animals don't necessarily mate with just any old individual that comes along. In addition to stimulating mating behavior in general, these odiferous pick-up lines can be crucial for advertising one's superior qualities during courtship.

UP CLOSE AND PERSONAL

I am often asked how we know whether birds are actually paying any attention to each other's odors. Unlike mammals, birds do not scent-mark. Also, most birds don't seem to have any obvious sniffing behaviors that we can see.

A notable exception is the pleasantly tangerine-scented crested auklet (*Aethia cristatella*), mentioned in Chapter 1. The origin of the auklets' odor is not entirely clear: auklet preen oil alone does not smell citrusy, and the odor is most potent in the feathers covering the nape of the neck in both sexes. Using behavioral tests, researcher Julie Hagelin has shown that crested auklets are attracted to this scent, and she suggests that it may be important for social or mating behavior. Indeed, during courtship in the breeding season, when the scent is strongest, both male and female crested auklets perform a ruff sniff, burying their bills in the nape feathers of their courtship partner, presumably to get a snootful of the tangerine perfume.

In my behavioral trials exploring the territorial responses of wild juncos, I noticed that birds defending their turf frequently stopped to wipe their bills on a branch between aggressive attacks. Birds often wipe their bills to clean them after eating, especially if they've eaten something they found distasteful, like a poisonous caterpillar, but taking time during an antagonistic encounter to clean your face didn't seem like a smart choice to me.

Earlier observers of bill-wiping during social encounters had dismissed this behavior as a displacement activity. In 1952, Niko Tinbergen, one of the founders of modern ethology,* defined displacement activity as an action that (1) is similar to a motor pattern that is normal for the species, (2) is seemingly out of context with the behavior preceding or following it, and (3) seems to appear "when an activated drive is denied discharge through its own consummatory acts"—in other words, when the animal is apparently conflicted or frustrated. When stressed, humans often show displacement activities, such as scratching your arms, twisting jewelry, or picking cat hair off your clothes during an uncomfortable conversation.

However, many animal behaviors that had been dismissed as displacement activities by researchers in the mid-twentieth century were later found to have adaptive functions. Just because the researcher does not yet understand why an animal is doing

* The study of animal behavior, from the Greek ethos, "character."

something does not mean that the behavior is irrelevant. For example, to avoid a fight with another male, a male monkey might suddenly decide that now is a good time to pick up and carry around an infant. This behavior is not parental care, but rather it is the male's attempt to shield himself from aggression during a potentially dangerous moment.

Could bill-wiping serve a purpose in aggressive contexts? While preening, birds wipe their bills against their uropygial gland to gather some preen oil, which they can then apply to their feathers. Presumably, there would be some preen oil residue left on the bill; the oil would be waxy and more solid after it cooled down from the gland's temperature. Were these birds wiping their preen oil residue on the branch in a heretofore undiscovered avian form of scent-marking?

It seemed like a wacky idea, and was a complete shot in the dark, but I decided to investigate anyway. If science has taught us anything, it's that not all wacky ideas are a waste of time, after all. (Witness: human evolution, dark matter, and heliocentricity.) The next summer I was planning to do fieldwork with my collaborator and friend Dustin Reichard (then a graduate student at Indiana University, now faculty at Ohio Wesleyan) in the beautiful Grand Teton mountains of Wyoming. This area is home to the pink-sided subspecies (*Junco hyemalis mearnsi*), a lovely little variant of the dark-eyed junco. I decided to conduct simulated territorial intrusions (STIs) with caged males as well as caged females to compare the target males' aggressive and courtship responses to the intruder.* I would record their responses, as was normal for these kinds of trials, but in addition to the usual data collection, I also kept track of how many bill-wipes the target male performed.

We got up at dawn every morning and drove out to that day's selected field site. My field assistant and I would hike along one of the

* We liked to call the trials with caged females "simulated sexy intrusions," or SSIs.

trails until we found an open spot in the woods. There, we would set up our equipment: our caged lure next to a speaker, an iPod connected to that speaker by a fifty-foot cable, and a video camera. We would sit with the camera and the iPod far enough away from the cage that we wouldn't interfere with the wild junco's behavior, but close enough that we could see—usually at least twenty to thirty feet. Then, we'd hunker down and play our recorded junco song. If we were lucky, we would be rewarded with a visit from a perturbed resident junco—and not interrupted by a tourist wanting to know what we were looking at. (Or worse, as on one terrifying occasion when a moose and her baby nonchalantly traipsed right through the middle of our trial. Moose mothers are notoriously intolerant of strangers near their little ones.)

I found that the male juncos in my study performed bill-wipes in response to both male and female intruders. But they did it *far* more often while courting a female (an average of three times during the five-minute trial, up to a max of about twenty times) than they did while threatening a male (an average of only 0.5 times per five-minute trial, maximum five times). Within courtship trials, the number of bill-wipes correlated with the time the males spent singing courtship song, the number of times they spread their tail feathers (to better show off their tail white), and the time they spent near the female's cage. In other words, the more work they were putting into courting the female, the more they bill-wiped. It seemed to be an important part of courtship that no one had previously noted.

What are they doing? I suspect that they are making an olfactory display. Just like a visual display shows off tail feathers or other attractive visual traits, an olfactory display should show off attractive smells. Rubbing the residual preen oils from their bills onto a branch or other material could release the volatiles trapped in the waxy oil through friction—much like lighting a scented candle releases the scents in the wax, or crushing herbs in your hands releases their aroma. Mammals do something like this too: prairie voles and gerbils groom themselves more frequently when they are around a conspecific, or even when just presented with one's odor. This behavior has been described as a kind of scent-marking, as it releases chemical cues into the air while the animal puffs up and licks its fur.

In my study, the males who did the most bill-wiping at females were the youngest males. Males in their first adult breeding season bill-wiped an average of six times per trial, compared to just once per trial by most of the males two years or older. However, these young males did not perform any of the other courtship behaviors more frequently than older males, so this difference was not just because the young males were courting more intensely. Perhaps the younger males were investing more energy into this olfactory display, using it as a way to signal their good qualities and to entice the females to overlook their lack of experience.

So, it seems birds do have behaviors that could function to make others notice their smell, particularly at close range. Once the two birds are close to each other, they can then use the many cues available to them—visual, chemical, and auditory—to determine whether the other would be a good mate. Chemical cues are likely very useful indicators of quality because they are tied to so many genetic, hormonal, and immune characteristics, and they can even relate to the animal's dominance within the social group.

THE SMELL OF QUALITY

Sometimes, it is helpful to break out of your comfort zone and study a different animal, in order to approach your research question from a new direction. In 2016, I had the opportunity to study a bird in which, unlike juncos, only a small percentage of males successfully woo females.

The lance-tailed manakin (*Chiroxiphia lanceolata*) is a small, colorful songbird in Latin America. Dr. Emily DuVal has been studying these birds on Isla Boca Brava in the Republic of Panama since 1999. I first met DuVal at the 2015 Animal Behavior Society conference in Anchorage, Alaska, where we were both judging an undergraduate research poster competition one evening. After the results were tallied and our job was done, the two of us chatted on our way back to our rooms, introducing ourselves and our research.

"Oh! I've seen your work, and I was planning to look for you at this conference," she said to me when I mentioned my recent focus. "I'd really like to measure preen oil odor in manakins, but I'm not sure how to get started." DuVal invited me to come to Panama the following field season to teach her and her crew how to collect preen oil. I agreed right away—I love travel and will go anywhere at the slightest invitation.

Historically, very few researchers have studied chemical ecology in tropical birds. Ornithology is dominated by studies of North American and European birds in temperate climates—likely because of the abundance of North American and European ornithologists—and many scientists make the mistake of drawing conclusions about the biology of all birds based on this limited sample. Animals living in tropical climates often evolve very different lifestyles compared to those that live in environments with seasonal changes in day length and temperature. Year-round warmth and rain lead to higher food availability as well as more pests and diseases, resulting in longer breeding seasons and different selection pressures, compared to temperate climates. I was excited for the opportunity to spend ten days in Panama catching and sampling manakins with DuVal.

Many of the sixty or so species of manakins have evolved complicated courtship ritual displays that require the cooperation of two or more males. These birds use a lek mating system to attract potential mates. In a lek, males gather together to display all at once, competing for the attention of females. Females use this comparison-shopping opportunity to observe all of the displaying males and choose which one to mate with. Since the male's display is often linked to his health or other aspects of his quality, females appear to use the information conveyed by the display to find the highest-quality male available to them. In most lekking species, about 10% to 20% of the males achieve 70% to 80% of the copulations. Once the mating occurs, the male's involvement in reproduction is essentially done. The female raises her offspring alone, so any benefits that she may receive from mating with a high-quality male can only be direct, such as the superior genes that her offspring will inherit, rather than indirect benefits, such as paternal care or territorial protection.

Lance-tailed manakin displays require two adult males, an alpha male and a beta male partner, who together perform a complex dance for discerning female onlookers. The two birds form a partnership that can last for years. Unlike similar partnerships in other species, these two males are usually unrelated. Adult males have a striking appearance, with black plumage, a bright red cap, and a patch of sky blue across the back that resembles a cape. There is no obvious visual difference between the alpha and beta males. In fact, the only way to tell which one is the alpha is to watch and see which one mates with the female. Together, these two males occupy a territory that includes a display perch—a branch of a tree or bush, usually low to the ground, where they perform their dance. This complex display includes several elements, beginning with coordinated calls while flying above the olive-green female who is waiting on the dance perch.

The most dramatic element of the ritual is the backward leap-frog dance. The first male leaps up from the branch and hovers in the air for a moment before descending and landing behind the second male. While the first male is hovering, the second male moves forward along the branch to the first male's vacated spot. The second male then repeats the first male's movements, leaping and hovering as the first male lands. They repeat these movements

Two lance-tailed manakin males performing a display.
Photo by Emily H. DuVal

for several minutes, making a distinctive call every time they leap. DuVal describes this call as "nraawn-raawnraaw," which is pretty much exactly what it sounds like.

In a classical lek, seen in species such as the sage grouse, all of the males gather in a single open arena when preparing to seek a mate. Lance-tailed manakin displays, on the other hand, are part of an exploded lek, which can be heard but not seen at the same time. Males seek out their own display territories, maintaining a display branch and keeping it clean for their dance. The dense foliage of the tropical rain forest means that the displaying males are visually obscured from each other, and females have to travel around to see them.

In the breeding season, which lasts from March through July, males practice their displays for hours each day. By the end of the mating season, alpha males are in much worse condition than other males. After putting all of their energy and time into displaying, often forgetting to eat and preen, alpha males weigh less, their feathers are ungroomed, and they have more parasites than beta males.

But this work often pays off: alpha male lance-tailed manakins sire nearly all of the offspring. One year, DuVal found that 100% of the copulations observed in her study population were performed by seventeen alpha males, out of forty-four alpha males and fifty-two beta males that they were tracking. Of course, there was the possibility that the beta males were mating when the researchers weren't watching. So, DuVal's team used DNA paternity tests and found that two beta males did actually sire chicks that year. However, when comparing paternity results across seven years, she found that fewer than 1% of chicks were sired by beta males.

Although alpha males are far more likely than beta males to reproduce, many more alpha males do not manage it at all. Achieving alpha male status is not a guarantee of success. What characteristics of alpha males are attractive to females, and what makes the males more likely to be successful?

DuVal found that females prefer older males who had been alphas for a longer time. Older males had more experience at being alphas and performing the display, and males who had the same beta partner for a long time could exhibit more coordinated dances. This result makes intuitive sense: a male who has held his alpha spot for a long time is likely a high-quality male with good competitive abilities.

One frequently studied measure of quality is the diversity, or heterozygosity, of one's genome. Heterozygosity—meaning that the two paired chromosomes, one from each parent, each carry a different variant of a gene—is associated with better overall health. Many detrimental genes are only bad for you when you have two copies of them. Having one functional copy of the gene is sufficient for the body's needs, and that second, inferior copy will then only be a problem if it is passed on to offspring who don't get a functional copy from the other parent. Being *heterozygous* also indicates that there is less inbreeding in your background: relatives are genetically similar and are more likely to produce *homozygous* (having two identical copies of a gene, one from each parent) offspring if they mate with each other. Thus, choosing a more heterozygous mate is thought to be an adaptive behavior, since that mate is more likely to be healthy and pass on quality genes to your offspring.

DuVal's research group tested whether reproductive success in alpha males was related to their heterozygosity. The researchers genotyped the males in their study at several neutral genetic markers called *microsatellites* (the same markers used in DNA paternity tests). These small, repeating stretches of DNA are commonly used as a proxy for measuring an individual's overall genetic diversity. Because they are nonfunctional and do not produce any proteins or regulate any other genes, they are free to mutate more often than functional genes, resulting in high variation across individuals. In contrast, functional genes—those that perform a biological task like coding for a protein—usually don't vary much in a population. Mutations can render functional genes defective, and the individuals who carry such mutated genes are less likely to survive. Thus, we use microsatellites or other highly variable nonfunctional genes to understand patterns of relatedness (such as paternity) and genetic variation.

The male lance-tailed manakins that sired the most offspring were indeed more heterozygous at these markers. Although this finding was exactly as hypothesized, such a result is not always guaranteed—sometimes we don't find statistically significant results due to small sample sizes or using inappropriate genetic markers, and sometimes the animals are behaving differently for reasons we do not yet understand. But this exciting result led to more questions. How could females know which males were more genetically diverse? They did not seem to differ in size, feather coloration, or behavior. DuVal and I decided to find out whether the information could be detected in their scent.

We collected preen oil from twenty-five males, including both alpha and beta males, and brought the samples back to the United States to analyze the volatile content with GC-MS. As it happened, there was no detectable difference between the scents of alpha males and beta males. Age did not seem to affect scent either. So, females likely determine the status, age, and experience of males based on their behavior, like their skill in performing the court-ship rituals, or just from familiarity with them over several breed-ing seasons.

However, we found a significant relationship between hetero-zygosity and preen oil volatile compounds. Specifically, we found that genetic diversity was negatively correlated with the concen-tration of several volatile compounds, including 1-hexadecanol and 1-heptadecanol. Males with low heterozygosity had more of these compounds in their scent, while males with high hetero-zygosity had less. Although we did not know for sure how high heterozygosity might result in lower amounts of these compounds and vice versa, we were thrilled to find that this crucial trait, in-visible to us, could be detected through scent.

These exciting results have led us to think even more about scent and courtship in the lance-tailed manakins and other lekking birds. For example, the males beat their wings very quickly while hovering during the backward leapfrog dance. Is this action help-ing to waft the males' scent toward their female audience? Perhaps these elaborate courtship rituals that we see in many bird spe-cies have important components besides the visual aspects that we readily observe.

We also found an intriguing relationship between scent and the time of year. As the breeding season progresses, male volatile blends shift, with increases in some compounds and decreases in others. As the breeding season winds down and the rainy sea-son gets nearer, mosquitoes are much more abundant. Interest-ingly, two compounds that increase in male preen oil at this time, 1-tridecanol and 2-pentadecanone, have both been shown to be effective mosquito repellents. Could these scented compounds be involved in repelling mosquitoes, and possibly reducing the likelihood of malaria infection? It would be fascinating to explore whether manakins carry around their own built-in citronella candle, but that's a question for another day.

Our study did not include very many males who successfully reproduced. In fact, only three males out of our sample of twenty-five sired any offspring that season, making it impossible to know whether the scent of heterozygosity really resulted in success. To get a better sense of how variation in odors truly affects reproductive success, I had to examine a larger dataset by going back to birds in which more of the males in the population were able to successfully reproduce.

THE SMELL OF COMPATIBILITY

"Good genes" might make a potential mate seem sexy in a number of ways. High levels of heterozygosity throughout the whole genome certify that the individual is not inbred and is not likely to be carrying two copies of any particular detrimental gene. This concept of good genes focuses on genes that are always good in any context. However, sometimes whether one mate's genes are "good" depends on the other mate's genes, and whether their combination will lead to favorable results in the offspring. This idea focuses on compatibility—that is, the two mates carry different copies of genes, ensuring that their offspring will have a mixture and thus be heterozygous.

Many studies have used microsatellite markers as a proxy for whole-genome diversity, just as we did with lance-tailed manakins. While DuVal concluded that females preferred more heterozygous males, other studies have instead found that some females prefer to mate with not just genetically diverse males but with those males that have different microsatellite genotypes than their own. Sedge warblers that migrate across the Sahara and the house finches commonly seen at backyard feeders in North America are among these species. But how do animals know if a potential mate has different genotypes from their own? While microsatellites are a useful proxy for researchers to assess whole-genome diversity, it is unlikely that animals have a way to detect nonfunctional genes. They are also probably not able to assess the entire genome of potential mates.

The easiest way to increase your chances of finding a mate genetically different from you is to simply avoid mating with relatives. Parent-offspring pairs are, on average, 50% genetically identical, as are full siblings. Half siblings share about 25% of their genome, and first cousins are about 12.5% similar.

Most animals, including birds, appear to be able to distinguish between close relatives and nonrelatives using scent. Learning and memory during early life are thought to be important in kin recognition, which can be accomplished through direct familiarization with close family members or can be acquired through more indirect means. In the latter case, animals could form a "template" in their memory of what close relatives smell like and subsequently compare the scent of unfamiliar animals to that template. If unfamiliar animals match the template closely enough, then they are likely related. Another mechanism, particularly important in solitary species, is that animals could compare their own scent to that of others—again, a match would indicate relatedness.

In behavioral trials using the classic Y-maze setup, European storm petrels strongly preferred the scent of unrelated individuals compared to the scent of very close relatives (either a sibling or a parent). Humboldt penguins at the Brookfield Zoo were able to distinguish between the scent of kin and non-kin (the relatedness of the kin used in the study ranged from first cousin to sibling or parent levels of similarity). Like the petrels, the penguins also preferred the scent of nonrelatives, even when both scents were from individuals they had never encountered before.

Zebra finch chicks can recognize the scent of their nest. In fact, they remember the scent of the nest they hatched in, even if they are then moved to a different nest and raised by foster parents. German researcher Tobias Krause, along with his mentor Dr. Barbara Caspers, conducted an experiment in which they moved newly hatched zebra finch chicks, between one and four days old, into a foster nest. Then, about three weeks later, they tested whether the chicks preferred the scent of their hatching nest or their foster nest. Nearly all of the birds chose their original hatching nest, showing that they could recognize the scent of genetic relatives and were not simply choosing the more familiar nest odor. Even more intriguing, in a later collaboration with Julie Hagelin and others, Krause and Caspers moved the chicks to foster nests before they hatched—and found that even though they had never met their genetic mother, the chicks could still recognize her scent.

Does the ability to recognize the scent of relatives actually affect mating decisions? To answer this question, Barbara Caspers set up triplet groups of zebra finches that included one female and two males—one completely unrelated and the other an unfamiliar full brother. If the scent of a relative affected mate choice, the researchers reasoned, then females with a functioning sense of smell should avoid mating with their brothers, but females who couldn't smell should be equally likely to mate with either male. Using a wash similar to a nasal spray, they applied a zinc sulfate solution to the olfactory mucosa of half of the females, which interfered with their olfactory neurons, making them anosmic for a few weeks. The remaining females were in the control group and only received a saline wash in their nares. After they had a few days to recover, all the females were placed in individual cages with two males and two nest boxes, so that each of the males could build a nest. (In zebra finches, the male collects the material and constructs a rough nest, which the female then refines.) The researchers then observed the behavior and reproduction of these groups for several weeks.

As predicted, the females that were anosmic did not discriminate between kin and non-kin: they were equally likely to mate with the unrelated male or the brother. About half of the females in this anosmic group successfully raised chicks.

The researchers expected that the females in the control group, who still had the ability to smell, would show a preference for mating with the unrelated males. However, once again, the study did not exactly go according to plan. The females in this group were less likely to mate at all—only two of sixteen females produced any offspring, and only one successfully raised her clutch. Even more surprising: in almost half of the triplet groups with females who could still detect scent, males were severely injured due to aggression and had to be removed from the cage. Why did this happen in the smelling females' groups but not the groups with anosmic females? It was not known which birds attacked the males, but the researchers speculated that the females were the aggressors. The females were caged in an environment where they could not escape from an unwanted potential mate, so their response may have been to attack instead.

So, animals, including birds, can recognize kin through scent—but how? As mentioned before, they could be somehow comparing the scent to either the scents they grew up with or to their own scent. Why do relatives smell like each other? A popular hypothesis suggests scent is influenced by molecules produced by MHC genes, and since relatives have similar MHC genotypes, they also have similar smells.

As noted previously, MHC—the major histocompatibility complex—is a large family of immune genes that allows your immune system to recognize when invasive pathogens are present and to alert the rest of the immune system to attack the pathogens. It's thought that having more MHC variants in your genome is best: the more diverse your MHC proteins, the more potential pathogens your immune system can recognize.

If having a diverse MHC genotype means that an animal is healthier overall, then perhaps indirect cues of overall health can convey information about an animal's genotype. For example, producing pigmentation for colorful feather ornaments takes more energy and nutrition than dull plumage, and these feather ornaments often indicate the quality of the animal. There is some evidence that such ornaments do correlate with MHC diversity in some birds, such as the common yellowthroat, a small North American warbler. Male common yellowthroats have a bright yellow patch on the throat and black feathers across the eyes and down the side of the face, referred to as the mask. Researchers noted that females preferred to mate with males with large masks— in particular, they often chose large-masked males for extra-pair mates. Peter Dunn and Linda Whittingham at the University of Wisconsin–Milwaukee, along with then-postdoc Jennifer Bollmer and collaborator Corey Freeman-Gallant, found that males with larger masks had more variable MHC genes. Furthermore, the males with more variety in their MHC genes were also more likely to survive to the next year, and males that had a particular MHC variant were more resistant to avian malaria.

Several examples of such correlation between MHC genotype and other aspects of a bird's appearance or performance have been identified, but there is a wide range of variation. One study revealed that MHC diversity is correlated with the complexity of male song sparrow songs. Ring-necked pheasant males have long spurs on their feet, which they use in fighting. The length of their spurs has also proven to be related to their overall MHC variation. Female song sparrows prefer males with complex songs, and female ring-necked pheasants prefer males with longer spurs. These examples suggest that female birds can choose a male with a diverse MHC genotype through indirect information. However, so far researchers have been unable to identify such clear correlations with visual or auditory traits in other bird species.

In fact, not all animals are choosing mates with the most diverse MHC. Instead, as predicted by the view of "good genes" as compatibility, some females prefer not the males with the most diverse MHC but instead those that have dissimilar genotypes from their own. This kind of mate choice, called *disassortative mating*, results in a different optimal mate for each female, and there is no one trait, like brighter feathers or larger size or more complex songs, that all females could use to detect it. Blue petrels are one such example. Researchers compared the variation at MHC genes in a population of blue petrels, and they found that mated pairs were more dissimilar at these genes than would be expected if they had randomly chosen a mate from the population. Sometimes the preference for MHC dissimilarity is not so obvious. In red junglefowl, males did not seem to care whether females were MHC-similar or dissimilar, and they mated with both. However, the males gave more sperm to the dissimilar females—a fact that researchers discovered by fitting a harness over the females' *cloacae* to collect the ejaculate, and then by counting the number of sperm.

In specially bred laboratory mice that are genetically identical *except* for their MHC genotype, mice are able to differentiate between MHC-similar and MHC-different individuals on the basis of scent—specifically the scent of urine. MHC protein products end up in mouse urine and ultimately affect the odor of the urine. Similarly, in the three-spined stickleback (a small fish used in many evolutionary and behavioral studies), small proteins (*peptides*) produced directly by the MHC genes affect the fish's odor. Female sticklebacks' preference for male odor can be affected by experimentally changing the peptides in the water to mimic the odor of a male that is MHC-similar or dissimilar to the female. At least in these species, there seems to be a direct link between MHC genotype and odor.

Since odor is the most likely way that MHC diversity or similarity is detected, preen oil volatile compounds are the most probable candidates for communicating this information in birds. Only recently have people begun to look for a relationship between preen oil compounds and MHC genotype—and they have found it. For example, as noted earlier, the degree of chemical dissimilarity between the preen oil odor of two song sparrows is correlated with the genetic distance between their MHC genotypes, though only when looking at the distance between a male and a female song sparrow. If the two birds being compared were of the same sex, there was no relationship, which suggests that this chemical information may only be important when assessing potential mates. Additionally, in a classic Y-maze preference test, both male and female song sparrows preferred the scent of MHC-dissimilar opposite-sex birds compared to that of birds with similar MHC genotypes. They also preferred the scent of more MHC-diverse birds.

Relatedness and MHC similarity is reflected in scent, but the preferences for those scents are not always what we would predict. What birds consider "sexy" could simply be in the nose of the beholder: maybe different males smell attractive to different females, depending on the female's own genotype. But there are many cases where most females prefer the same few males, as in lekking birds like the lance-tailed manakin. And in other birds, we can clearly distinguish certain traits that more attractive and successful males have, like plumage or song. Clearly, there is no single universal answer to the question of how mate choice works.

THE SMELL OF SUCCESS

Pheromone researcher Dr. Tristram Wyatt referred to the evolution of reproductive chemical signals as "success of the smelliest." Charles Darwin also acknowledged that chemical signals could be one of the outcomes of sexual selection, if females were most attracted to "the most odiferous males."

I decided to address the question of which odors female birds find attractive by measuring the scents of the males that they actually mate with. Up until that point, most studies of odor and odor preference in birds had either focused on preference tests in controlled conditions or relied on statistical correlations between odors and other traits that relate to quality or compatibility, such as hormone levels, aggression, relatedness, or genetic diversity. I wanted to compare more and less successful males in a wild population where many males were able to mate, not just the highest-quality birds.

Fortunately, I already had a large dataset of volatile profiles from dark-eyed juncos at Mountain Lake Biological Station. Back in the summer of 2008, I had studied how changes in testosterone levels correlated with changes in odor in this population by sampling preen oil from all of the birds we caught in the annual census and looking at the change in odor over time. Now, to understand how scent related to reproductive success, I would need data from pairs of juncos that had raised and fledged offspring. Not all breeding adults are successful, of course. Successful pairs must hatch healthy nestlings with no genetic defects and keep them disease-free, well fed, and warmed, and they must also protect their nests from brood parasites (like brown-headed cowbirds) and, especially, their abundant predators. Predators are everywhere, and the biggest threats to junco nests are snakes, chipmunks, and predatory birds.

For the 2008 season, I had volatile compound data from sixteen females and thirty-five males. How many of them had successfully reproduced? Dr. Nicole Gerlach (now a faculty member at University of Florida) had dedicated much of her time as a graduate student to running paternity tests, using blood samples collected over many field seasons, to understand how often and why females cheated on their mates. I asked Nicki if she could compare my list of birds to the paternity data from that year. We were delighted to discover that we had reproductive success and paternity data for twelve of the sixteen females and twenty-two of the thirty-five males.

I had collected my preen oil samples during the early season, shortly after the birds migrated back to the area and before they began building nests and laying eggs. If preen oil odor is important for attracting mates, this time is exactly when the birds should be producing their most attractive scents. Was their smell at this time related to the number of offspring the birds successfully fledged? Because about 25% of chicks in this population are sired by extra-pair males, I could look at "success" from a few different angles. How many surviving offspring did a male genetically sire (including his own nest and any extra-pair offspring he may have had)? How many offspring did he successfully help raise in his own nest—in other words, was he a good dad? How many of the offspring that he raised were sired by a different male?

I organized the data in a spreadsheet and ran a simple statistical test to see if there were any relationships worth pursuing. To my extreme surprise, that first test revealed a strong correlation between junco odor measurements and the number of offspring they produced. I'm never right on the first try, so this occasion felt like a huge victory. I discovered that the males who had sired the most genetic offspring had more of three particular volatile compounds: the methyl ketones 2-tridecanone, 2-tetradecanone, and 2-pentadecanone. Interestingly, I found the exact opposite relationship in females. Females with lower levels of these compounds had more offspring. It puzzled me that the sexes would have opposite patterns. I knew from my earlier work that breeding males typically have a much higher concentration of these compounds than breeding females do, although even within males there is variation. So, one way to look at this finding was that males who smelled more "male-like" (in other words, had the highest amounts of these compounds) had more offspring, while females who smelled more "female-like" had more offspring.

Those results tell us something about what birds might look for in a mate in terms of whose genes would be best to pass down

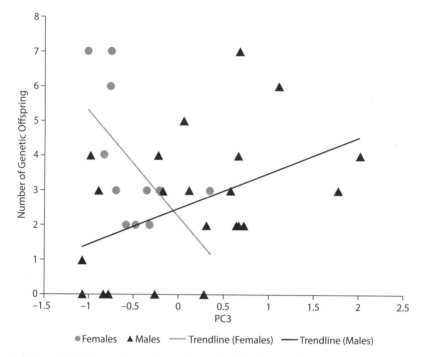

●Females   ▲Males   ——Trendline (Females)   ——Trendline (Males)

Relationship between the number of genetic offspring produced by dark-eyed juncos and a measure of the volatile compounds in their preen oil. Principal component 3 (PC3) is a composite score based on measurements of specific volatiles that correlate with each other. *Adapted from* Whittaker et al. 2013, *Animal Behaviour* 86:697–703

to their offspring. But what about parental ability? Since juncos engage in extra-pair copulations fairly frequently, perhaps some females choose a social partner who would be a good dad but then seek a different male to actually contribute the genes. I wondered whether information about one's parental ability is communicated via scent. I compared the scents of males who successfully raised offspring—just their own, their own plus those sired by an extra-pair male, or in a few cases, only extra-pair offspring. The males who successfully fledged the most offspring had a significantly different odor, with a greater concentration of the three "male-like" methyl ketones listed above and of four linear alcohols that are known to increase in both sexes during the breeding season. Again, since these preen oil samples were taken at the very beginning of the spring, before any breeding had begun, I concluded that females may be able to predict which male would make the best father based on scent alone.

Does a male's scent predict whether his partner will cheat on him? Using genetic paternity tests, I calculated how many of the nestlings in a male's nest were not his own offspring, and then I looked for a correlation with scent. I found that males who had higher amounts of dodecanoic, tetradecanoic, and hexadecanoic acid had more extra-pair offspring. Interestingly, these three compounds are sexually dimorphic—but they are usually found in higher amounts in *females*. It seemed that females were less faithful to males who smelled more female-like.

These results don't tell us if females used the information present in odor to choose their mates, or if odor influenced whether they mated with an extra-pair male. But they do show that preen-oil-based odor predicts an individual's potential quality and ability to produce and raise offspring. Testing whether odor alone influences such behavior would require conducting highly controlled experiments, with birds whose odors have been manipulated, and allowing the birds to mate and produce offspring.

I haven't done this work, nor do I currently plan to, mostly because I lack the necessary facilities and resources. I hope that someone does carry out such experiments, though. I don't have a traditional tenure-track faculty position, and I don't have my own laboratory or field site. So, I have to design experiments that are realistically within my reach.

CHAPTER 5

MAKING SCENTS

OF BACTERIA

The purely theoretical process is managed by the tools of the senses of seeing and hearing; what we see or hear we leave as it is. On the other hand, the organs of smell and taste are already the beginnings of a practical relation. For we can smell only what is in the process of wasting away, and we can taste only by destroying.
    —GEORG WILHELM FRIEDRICH HEGEL,
      Lectures on Aesthetics

SNIFFING OUT A CAREER PATH

It's not that I didn't *want* to follow a "normal" academic career path. These days, tenure-track jobs are getting increasingly difficult to come by. Universities continue to produce PhDs, but they contract more teaching positions to contingent faculty rather than create new tenure-track openings. Still, I was determined to try. During my four years as a postdoctoral researcher at Indiana University, I applied to literally hundreds of tenure-track jobs at all different kinds of colleges and universities: big state universities, small liberal arts colleges, private universities, community colleges, small branch campuses. Finally, in 2010, I landed four on-campus interviews.

All four institutions that were interested in meeting with me are categorized as "comprehensive" universities—not quite big research institutions like the flagship state universities and not quite liberal arts colleges. Faculty at these colleges typically have a heavy teaching load, yet it was becoming more common for their ambitious administrations to impose research expectations on the faculty, despite not providing sufficient time or resources. Although all of the faculty I met during my interviews loved both teaching and research, they were stretched beyond their limits. Everywhere, it seemed, faculty were overworked and underappreciated. No one seemed happy.

From these interviews, I received a single job offer, at Indiana University at Kokomo. IUK is a regional campus in the IU system, located in a small town about an hour north of Indianapolis. I had good experiences with IU during my postdoc in Bloomington and a network of friends in Indiana and the Midwest in general, so I was generally positive about the possibility of staying nearby. I had enjoyed meeting the faculty and staff during my interview there, and the teaching opportunities presented by the position were well within my areas of interest. However, the university administration had made it clear that they were not interested in supporting my research with any facilities or funds. I had dreamed of having my own small aviary where I could keep captive birds and conduct studies on campus, enabling me to provide research opportunities for my undergraduate students. I was reluctant to let that dream go, but I was also hesitant to turn down a tenure-track job offer in the increasingly tough job market. They demanded my answer within two days, which was a bit of a shock. Two weeks is a more standard time frame for considering this kind of offer. I responded by expressing my continued interest and asking for the usual two weeks to make my decision. I had not yet heard from the other three universities, and two weeks would give me time to contact them and ask if my candidacy was still under consideration. The provost responded by rescinding the offer.

This version of the story was not the one that reached the IUK faculty. I received emails from faculty members that I had met with, expressing disappointment that I had turned them down and accusing me of making unreasonable demands. Some assumed that I had been looking for a spousal appointment for my husband (which would have been nice but not expected) and angrily informed me that their college didn't have those kinds of resources. I didn't bother responding to any of these emails to set the record straight. What was the point? I was young and needed a job. Starting trouble couldn't possibly benefit me.

By the end of that spring, I had racked up a pile of job rejections and had lost any desire to keep looking for a traditional faculty position. I had seen misery everywhere, and all I had experienced was discouragement—sometimes even outright abuse. But I loved academia, and I loved doing research. I started looking around me for fresh inspiration.

The tenure-track faculty search is a well-worn road, with standard processes and resources to find listings. There are differences between academic fields, of course, but in general you can find excellent advice from people who have been through it all before. The so-called alternative academic path, however, feels more like wandering in thick jungle with nothing more than your wits and a machete, trying to determine which mass of vegetation to hack through.

Most large universities have central laboratory facilities and centers for research featuring expensive equipment, such as DNA sequencers or electron microscopes. Such centers are available for use by researchers throughout the university. Around the same time that I was coming to grips with the idea of giving up on the tenure-track dream, I was working on a transcriptome sequencing project that involved IU's Center for Genomics and Bioinformatics. I had a few research meetings with the then-director of the center, John Colbourne, and soon realized that his was not a typical faculty position. I asked questions about his research and his job, and I learned that he was a researcher with strong academic credentials, that his primary responsibility was running this center, and that he was able to conduct his own research while managing the lab. And he didn't have to teach. This sounded like a perfect job to me. The puzzle, of course, was how to find such a position.

I was incredibly lucky—I didn't have to wait long for my opportunity. I subscribed to a popular evolutionary biology email list, and I regularly received emails about new openings. Most of them, but not all, were faculty positions. One day, a golden opportunity appeared. A brand-new center at Michigan State University was hiring a managing director.

In 2010, the BEACON Center for the Study of Evolution in Action had just been selected as one of the newest Science and Technology Centers (STCs) funded by the National Science Foundation. Compared to most NSF grants, the STC grants are large scale and long term: $5 million per year for ten years. The aim of the STC program is to support large interdisciplinary research and education projects involving collaboration across multiple universities. According to NSF, "STCs focus on creating new scientific paradigms, establishing entirely new scientific disciplines and developing transformative technologies which have the potential for broad scientific or societal impact."

The managing director's role would include participating in policy formulation and planning of center activities, managing the internal funding process, preparing reports to NSF documenting the center's activities, organizing conferences and workshops, and—yes!—engaging in research relevant to the center.

As the manager for the Ketterson lab, I had acquired quite a bit of relevant administrative experience. I managed the lab's research permits, submitted annual reports, maintained our compliance with safety regulations, and oversaw many of the day-to-day operations. Although there were only ten or so people in the lab—far fewer than the hundreds involved at BEACON—everyone has to start somewhere. With the same vigor I had thrown into applying for the position at IU, I poured all of my energy (or what was left of it, anyway) into this one application.

It was due June 15, with a start date of August 1—a very short timeline. I submitted my application early. And then I waited. And then I tried to stop thinking about it. They probably had an internal candidate, I reasoned. It's possible that my thoughts on the matter were influenced by the rejection I had just received after my fourth and final on-campus interview. The position had gone to someone who had been teaching at that college already.*

But one afternoon, the BEACON Director, Dr. Erik Goodman, emailed me to acknowledge receipt of my application. That one email instantly ruined all hopes of productivity for the rest of the day, as I started obsessing about the possibilities of this position. Things began to move quickly after that. I was invited for a video-conference interview that was very different from every tenure-track interview I had participated in. Instead of meeting individually with faculty members, or giving a talk to a large audience, I met with a dozen people seated around a conference room table, all taking turns asking me questions. I had no familiarity with the business jargon often thrown around for management jobs and wasn't always sure what I was being asked. So, I settled for being open and honest about my working style and how I would run things.

Very shortly after that, the director invited me to Michigan State for an on-campus interview. Compared to the stressful ordeals

---

* Though we are quick to believe that internal candidates always get the job, according to science blogger Jeremy Fox, it's actually quite rare (https://dynamicecology.wordpress.com/2017/09/23/hardly-any-ecology-faculty-positions-are-filled-by-internal-candidates-and-you-cant-reliably-identify-the-ones-that-will-be/).

I had experienced for tenure-track jobs, this interview seemed surreal. Everyone was so nice. Of course, they were excited about receiving this huge grant, but more importantly, they were excited about working with each other. They genuinely seemed to enjoy their colleagues and their jobs. I had never experienced this atmosphere before in an interview, and I admit that my hard-earned cynicism made me a little suspicious.

Despite a few lingering doubts, I readily accepted the offer when it came and found myself figuring out how to run a multi-million-dollar research center nearly overnight. It was exhilarating and exhausting. My persistent case of imposter syndrome—the belief that you don't deserve your success, usually accompanied by the fear that someone will realize that you're a fake—was stronger than ever. But over time, I discovered that I was fully capable of doing this job successfully. I learned how to manage large projects like writing the annual report and planning the annual conference. I learned how to be a good administrator, helping people get what they needed to do their research. And within a year, I was able to make time for my own research.

BACTERIA AS CRAFT SCENT BREWERIES

My nontraditional appointment at Michigan State University came with no lab group and no built-in peer group of researchers working on the same general concepts as I was. I had no laboratory of my own, and I couldn't take on any graduate students to work with me. However, what I did have was a web of connections to literally hundreds of researchers at BEACON with a diverse array of skills, interests, and specialties. The possibilities for collaboration were overwhelming. But first I had to figure out what I was doing.

I used our database to look up all the other BEACON members at MSU who also studied chemical communication in animals. Although there weren't many, I did find a few researchers at various career stages studying all kinds of animals. With the help of neu-robiologist Dr. Heather Eisthen, who studies olfaction in amphibi-ans, I formed a group where we could meet semiregularly to share what we were working on and give each other feedback. We called it the Chemical Communication Group, or CCG.

At each of our CCG meetings, one person would give a presen-tation about a research project in progress, and we would discuss ways to approach or solve problems that person was encountering. I no longer remember exactly which project I talked about first, but I do remember one question at the end of my talk, because it changed my entire research trajectory.

"These compounds you've listed—all the linear alcohols, car-boxylic acids, etc.—are all known to be by-products of microbial metabolism," said Kevin Theis. Kevin—tall, dark, and stoic—was a postdoc at MSU at the time and studied scent-marking in spot-ted hyenas. "Have you looked at whether there are bacteria in the uropygial gland that might be producing these odors?"

No. I had never once considered that. At that point in my career, in fact, I hadn't thought much about bacteria at all. But Kevin had. Like many mammals, hyenas spend much time and energy mark-ing their territories. Right under a hyena's tail is a scent pouch that produces a thick, yellow-green paste, which the hyena rubs on stalks of grass. This dark, moist scent pouch is full of bacteria fer-menting the fatty secretions from the anal glands. Kevin's post-doctoral research focused on understanding the relationships be-tween these bacteria in hyena scent pouches and the odors given off by the scent-marks.

The idea known as the fermentation hypothesis for chemical recognition was first described in the 1970s, in the work of two different researchers: Edward Albone's studies of red fox scent se-cretions and Martyn Gorman's research on the Indian mongoose. Both of these scientists used antibiotics to experimentally remove bacteria from the animals' scent glands. As a result, the glands no longer produced short-chain fatty acids, which are important volatile compounds in mammalian scent-marks. The researchers concluded that bacteria living in the scent glands were responsible for producing these odors, and that if you change the bacteria, you change the odor. Thus, the two basic ideas of the fermentation hypothesis are (1) bacteria live in animal scent glands and contribute to the odors produced by these glands, and (2) variation in the animals' scents reflects variation in the bacterial communities on their bodies.

By the time I met Kevin in 2010, several studies had described the contributions that symbiotic bacteria make to the scent-marks of various mammals. Specialized scent glands aren't the only places on animal bodies where bacteria grow, of course—nearly every one of our body parts has its own associated symbiotic microbial communities. Bacteria feed on whatever is available to them in those places: for example, gut bacteria consume the food we ingest, bacteria in our eyes feed on our tears, and bacteria on our skin get nourishment from the components of our sweat and skin oil. Mammalian scent-marks can include urine, feces, the secretions from scent glands, or a combination of these substances.

Bacteria can even change the information present in proteins produced by an animal and used as scent-marks. When elephants are in breeding condition, the males go into a condition known as musth. An elephant in musth has elevated hormones, especially testosterone; its appetite decreases, and its energy is spent searching for potential mates and aggressively competing with other males for them. This condition also features nearly constant urine dribbling, so that the male leaves a trail of odiferous protein compounds that advertise information about the elephant's identity and health. Bacteria are also present in the urine, and over time the bacteria consume these proteins, slowly changing the odor given off. This interaction between the bacteria and the urinary proteins may result in a kind of timed release of chemical information, so that recipients can detect not only who the elephant is but how long ago he was in the area.

There are several known examples of how bacteria metabolize animals' biological products to produce volatile compounds that are used by the animals for communication. In the wild cavy, a South American relative of the domesticated guinea pig, males scent-mark using secretions from their perineal glands, so named because they are located in sacs just below the anus. The secretions from the gland collect in the sacs, where fermentative bacteria live. The perineal sacs also contain urine, probably due to the way the animals drag the sacs along the ground while scent-marking. The urine itself appears to be an independent source of odor, as male cavies are interested in the scent of the urine alone, in addition to the perineal secretions. When given the choice between pure secretions taken directly from the perineal gland and the "dirty" secretions that had fermented in the bacteria-containing sacs, male cavies were much more interested in the "dirty" scent. Researchers concluded that the fermentative bacteria within the perineal sacs, not the animals' own secretions, produce the biologically significant odor components of the scent-mark.

Components of urine can be processed and broken down by bacteria into compounds used for communication by animals. In the greater sac-winged bat, an insect-eating mammal found in Central and South America, males use scents from their wing pouches during courtship by hovering in front of females and fanning the odor at them. However, there are no scent-producing glands in the wing pouches. Instead, the males fill these sacs with secretions from other parts of their body. First, they orally transfer their own urine to the sacs. Next, they press the tip of their penis to the sacs, adding a sperm-free droplet to the mix. This behavior, called perfume blending, takes at least half an hour and is performed daily. Bacteria that colonize the sacs break down the secretions into the volatile compounds that make up the scent that females find so attractive. By cleaning and refilling the sacs daily, the bats control the degree to which the bacteria break down the components and degrade the odor.

The types of bacteria that inhabit various places on an animal's body, whether glandular sacs or skin, are not random. Animals acquire bacteria through physical contact with their environment—not just plants and soil but, importantly, other members of their own species. Animals have evolved many behaviors that help them collect the right kinds of bacteria from their physical and social environments. Coprophagy (from the Greek copros, "feces," and phagein, "to eat") is widespread across the animal kingdom and helps young animals colonize their guts with bacteria needed to digest complex starches.*

Living in a social group means having physical contact, which inevitably leads to spreading bacteria to your group members. This microbe sharing ultimately has the effect of creating a group-specific bacterial profile and odor. Kevin's work on spotted hyenas in Kenya demonstrates clearly how this can happen. Spotted hyenas live in large clans of about forty to eighty animals. Females are philopatric (from the Greek philos, "beloved," and patra, "homeland"), meaning that they stay in the clans where they are born. Females maintain a strict linear dominance hierarchy, in which higher ranking females have priority of access to resources like food. Daughters inherit their mothers' rank as soon as they are born, so the young daughter of a high-ranking female outranks a low-ranking adult female. Usually, the adult males in a group are all immigrants from other groups. This pattern ensures that males are unrelated to all of the females in the group, and thus they can easily avoid inbreeding. Hyenas defend their territory from

* My dog Eleanor is an enthusiastic poop-eater. I think it is disgusting and embarrassing. But when she's refusing to come in from the backyard because she's busy chowing down on feces left by my other dog, I tell myself that she's just getting the gut microbes necessary to keep her healthy.

neighboring hyena clans, regularly patrolling their territory and scent-marking the borders. The hyenas, especially the young ones, often engage in over-marking, scent-marking the same place that another animal has recently marked. Over-marking not only adds the hyena's own scent to the mark but also exposes its scent pouch to the bacteria left behind in the paste of previous markers. Kevin's work showed that the scent pouches of hyenas from the same clan had similar bacterial communities and similar scents. Furthermore, each clan had distinct, group-specific bacterial communities and scents.

Kevin also found differences between males and females, and between females in different reproductive stages. Female hyenas nurse their young for a year or longer, until the mothers become pregnant with their next litter, so most adult females are always either pregnant or lactating. Kevin found that the scent pouches of pregnant females were distinct from those of lactating females, likely related to physiological changes caused by these reproductive states.

An animal's own hormones can affect bacterially produced scents. For example, differences between male and female odors can arise through the bacteria's direct consumption of sex hormones. The secretions of the sebaceous (oil-producing) glands and sweat glands in the human armpit are initially odorless. However, human armpits are home to several bacterial species that consume these secretions and produce volatile odors as a by-product. One species of Corynebacterium in particular metabolizes precursors from these secretions, including androgens, producing a musky scent. As a result, people with higher androgen levels have a different odor compared to those with lower levels.

Not all bacteria that an animal encounters can colonize just any part of its body. Whether an animal's body is a welcoming environment for a particular microbe depends on the response of the animal's immune system, its hormone levels, and the diet that it consumes, among other factors. Thus, it is possible for scents produced by resident bacteria to indirectly reflect information about the host animal's biology.

When I first began learning about the role of bacteria in producing animal odors, nearly all of the research published had focused on mammals, though there was some work on insects, reptiles, and fish. But despite a growing body of research on feather-degrading bacteria and the antibacterial properties of preen oil, no one had yet investigated whether bacteria were important in avian chemical communication. After all, the myth of avian anosmia still held sway. It seemed high time that someone addressed this mysterious matter. Someone, in this case, meant me.

THE UROPYGIAL GLAND AS A HOST FOR BACTERIA

When Kevin Theis and I decided to work together to research odor-producing bacteria in preen oil, one of the first things I did was to literally look for it. I took a few samples of frozen preen oil to the Center for Advanced Microscopy on campus at Michigan State. Using a scanning electron microscope, the lab tech and I examined preen oil smeared on glass slides.

Preen oil is an odd, thick substance, and one that the lab tech had no experience with. We had a tough time determining whether any of the blobs we saw were actually bacteria. There were quite a few interesting structures that were probably fats and waxes. However, when I emailed some of the images to Kevin, he was confident that at least a few of our images were of bacteria. It was an encouraging start.

But what kind of bacteria lives in preen oil? Many of the previous studies on bird-associated bacteria were focused on harmful bacteria, especially feather-degrading varieties, such as *Bacillus*. These bacteria can break down the keratin in a bird's feathers, which may decrease flight efficiency, reduce insulation of the feathers, and even dull the feathers' color. Laboratory experimental work showed that preen oil applied to bacterial cultures in petri dishes inhibited the growth of these feather-degrading bacteria. In an interesting twist, researchers discovered subsequently that the source of preen oil's antimicrobial properties is secretions from other bacteria. An important set of studies was conducted on the Eurasian hoopoe, a cavity-nesting bird with striking black and white striped wings and a large striped crest on its head. During breeding season, the hoopoe's preen oil changes from the usual whitish color to a dark brown and becomes quite smelly. The cause of this change is a massive increase in the growth of *Enterococcus* bacteria in the uropygial gland. The females apply this smelly oil to the surface of their eggs, and it helps protect the embryonic chicks within from infection. Bacteria are both villains and heroes in this story.

Kevin and I wanted to learn about the composition of the whole bacterial community in the junco's uropygial gland. We might find helpful bacteria like the *Enterococcus* in hoopoes. And, more crucially to me, we might also find odor-producing bacteria. To get started in our exploration, I went to Mountain Lake and used sterile cotton swabs to sample the glands of a couple dozen juncos. To do so, I rubbed the swab back and forth across the gland, which stimulated it to secrete a little bit of preen oil. In this way, I was able to obtain bacteria from both the inside of the gland (secreted along with the oil) and the outside surface of it. Because birds take preen oil from their own glands in much the same way, using their beaks instead of a cotton swab, this method seemed like it would get the most representative sample of bacteria that come into contact with preen oil.

Shifting my focus to bacteria added a whole new layer of complexity to my fieldwork. I had to learn about sterile techniques to avoid contaminating the samples with bacteria from my own hands and other surfaces. Now, I had to wear gloves whenever handling the birds, which took some getting used to. It was harder to disentangle birds from nets while wearing gloves—I hadn't realized just how much I depended on the feel of their feathers to find the nylon threads wrapped around the birds' bodies. But when I suggested handling the birds bare-handed, taking care not to touch the uropygial gland, and then putting on gloves to take samples, Kevin was sternly disapproving. By this time, he had become more microbiologist than field biologist, and he was not sympathetic to my complaints.

The samples had to be stored in special freezers at -112°F, vastly colder than normal freezer temperatures of -4°F, which we used for storing blood and other biological samples. It was important to make sure that the bacteria did not degrade, so that we would be able to get accurate DNA sequences to identify them. When I first started collecting bacterial swabs, the only adequately cold storage at Mountain Lake Biological Station was an old chest freezer that didn't close properly on its own—the lid was weighted down with an elephant skull.* (Fortunately, the station purchased a new freezer the following year.)

Until fairly recently, the only way to find out what kind of bacteria were present in a biological sample was to culture them. You would simply swab the sample on a petri dish filled with agar (kind of a starchy gelatin which provides nutrients to the bacteria) and let the bacteria grow into colonies. The biggest disadvantage to this method was that the vast majority of bacteria in the world are difficult to culture under standard laboratory conditions, so only a comparatively small number of bacteria could be identified in this way. However, in our current age of so-called *next-generation sequencing* methods, we can instead target the DNA of all of the bacteria in a sample at once. These techniques take advantage of the fact that DNA copies itself under certain conditions. When

---

* I have no idea where the elephant skull came from, but it was probably from the biology teaching lab.

DNA synthesis is stimulated by chemical and thermal reactions, the replicating strands of DNA incorporate fluorescently labeled nucleotides (the "letters" in a strand of DNA) added to the mixture. The sequencing machine then detects these tagged molecules and can use this information to "read" the order, or sequence, of the DNA. We can then use the DNA sequence data to identify the bacteria.

To sequence and identify bacteria in a community, we used a standard method that targets a single slow-evolving RNA gene, 16S ribosomal RNA. Ribosomes are found inside cells, and they synthesize proteins by linking together amino acids in the order specified by the genes coding for them. Ribosomal RNA, or rRNA, is used by the cell in this process, and the 16S gene plays an important structural role in the ribosome. Like many structural genes that contribute to the building of cell parts, 16S mutates slowly from an evolutionary point of view, since most mutations are likely to harm the structure of the ribosome and render the cell nonfunctional. Thus, members of the same species usually all have identical or nearly identical 16S sequences, while a different species or genus will have its own unique mutations. Genes like 16S rRNA are helpful when trying to analyze how closely related different species are, and they can be used to identify members in a bacterial community. One way to think of it is as a DNA barcode, since it functions for scientists in much the same way that a grocery store barcode identifies the product when the cashier scans it. This method does not always allow identification all the way down to the species level, but we can usually identify the bacterial genus or at least family.

Kevin sequenced and analyzed the swab samples I collected from the Mountain Lake juncos. Junco uropygial glands, we came to find out, are home to unexpectedly rich and diverse bacterial communities. To our surprise, we identified far more types of bacteria in our samples than Kevin had found in hyena scent pouches. Many of the bacteria we identified had been found on bird feathers in other studies, including Enterococcus (the source of antimicrobial substances in hoopoe preen oil), Staphylococcus, and Pseudomonas.

Now it was time to answer my most pressing question: Did any of these bacteria produce odors? Specifically, did they produce the volatile compounds that made up junco odors? I looked up the most common bacteria that we found in "mVOC: A Database of Microbial Volatiles" (http://bioinformatics.charite.de/mvoc/). This online database compiles published information on volatile-producing bacteria, many of which are important in agricultural, medical, and biotechnical applications. I discovered that many of our bacteria were indeed known odor producers, including Enterococcus, Staphylococcus, Pseudomonas, Burkholderia, and Acinetobacter, along with several others. Best of all, we discovered that many of these bacteria produce the same volatile compounds that junco preen oil does, such as 2-undecanone, 2-tridecanone, dodecanoic acid, and tetradecanoic acid.

Not all of the bacteria in our samples were odor producers. Unlike a mammalian scent pouch, the secretions from the uropygial gland serve many functions for the bird, which may explain why these bacterial communities were so much more diverse than what Kevin found in hyenas. Since we know preen oil helps protect feathers from parasites and harmful bacteria, it was no surprise that in our samples we found several bacteria that produce antimicrobial compounds. Some, like Methylobacterium, Enterococcus, and Pseudomonas, produce chemicals that target other bacteria, while others, such as Arthrobacter and Burkholderia, make antifungal compounds.

It may seem counterintuitive that bacteria make antibacterial compounds, but this is one way that bacteria compete with each other. Microbes usually live in communities consisting of many different species, and they all compete for the same resources.

Those bacteria that can produce chemicals that harm other species, while not harming their own relatives, will be more successful than those that have to share more. These microbes can produce more offspring and grow a larger colony of their own relatives, which can result in cooperation and further dominance over the rest of the community.

Some bacteria that we found in the uropygial gland, including various types of *Burkholderia* and *Pseudomonas*, are used in agriculture for biocontrol, leveraging just this ability to kill other microbes. Others are able to digest petroleum and have been used in bioremediation to clean up pollutants. Undoubtedly, resident bacteria are helpful for birds in many ways. Symbiotic relationships, in which the bird provides resources and the bacteria provide protections, likely evolved first. The ability to recognize group members or relatives or potential mates using the odors that are a natural by-product of the bacteria's biological processes likely evolved later.

But all that big picture thinking about behavior and coevolution would have to wait until I could learn whether these bacteria in the uropygial gland were truly responsible for the juncos' chemical signals. What would happen to a bird's odor if the bacteria were taken away? Happily, I could treat this bacterial quandary the same way any doctor would—with antibiotics.

FINDING THE SOURCE

If you want to know for sure whether X causes Y, you need to take away X and see what happens to Y. To find out whether bacteria in the uropygial gland produce the volatile compounds given off by preen oil, we needed to eliminate the bacteria in the gland—or at least as much as we could. When researchers discovered that symbiotic bacteria were the source of antimicrobial compounds in hoopoe preen oil, they did so by injecting antibiotics into the hoopoes' uropygial glands. Then, they measured the antimicrobial properties of the resulting secretions, and they found that these properties were greatly reduced. We decided to take the same approach with our juncos.

First, we had to choose the right antibiotic and calculate the proper dose. Kevin and I consulted a veterinarian from the Potter Park Zoo in Lansing, Michigan, to make sure we got this part correct.* Based on the vet's advice, we chose a broad-spectrum antibiotic called enrofloxacin, often known by its brand name, Baytril. This antibiotic would kill as many different kinds of bacteria as possible, which was the goal in this project since we did not yet have enough information to try to target specific bacteria.

I borrowed twenty of the juncos in Ellen Ketterson's captive colony at the Kent Farm Bird Observatory at Indiana University for this experiment. Ten birds received an injection of enrofloxacin directly in their uropygial gland every day for five days. As an experimental control, the other ten birds also received injections, but their injections only contained a saline solution.

Antibiotics never wipe out all of the bacteria that are exposed to them, but they do target bacteria with certain characteristics and change the composition of the communities they affect. When you take antibiotics for an infection, for example, the drugs are designed to kill or slow the reproduction of the pathogenic bacteria, but they will also kill normal members of your bacterial community, including helpful ones. There are other bacteria in your gut that are unaffected by the antibiotic, and those will take advantage of the resources left behind by the killed-off bacteria, growing and reproducing and increasing in abundance, resulting in a very changed bacterial community. If you've ever been encouraged to consume probiotics, such as yogurt, when taking antibiotics,

---

* The best part of this visit to the zoo was that we got to meet with the vet behind the scenes in the keeper's area. While we were there, some of the other staff were trying to examine a penguin. The very cranky penguin did not want to get on the scale.

that's why—they help you seed a new community of beneficial bacteria that can compete with the less helpful left-behinds.

One of the first things I noticed about the birds in the antibiotic treatment group was that, after just a couple of days, their uropygial glands appeared to be drying up. Instead of their normal juicy appearance, they looked deflated and crusty, and some of them were difficult to sample at the end of the experiment. The glands of the saline group appeared normal. The antibiotics definitely seemed to be affecting preen oil production, but I wasn't sure how.

We compared the bacterial communities of all twenty birds at the beginning and at the end of the study. In the group that received antibiotics, the composition of the bacterial communities changed considerably, but the bacteria were not completely removed. The biggest change was a reduction in the abundance of two bacterial types: Staphylococcus and Pseudomonas. In response to knocking down these common bacteria in the glands, a few other types of bacteria increased in abundance, including Herbaspirillum and an unclassified species of Sphingomonadaceae.

Next, we compared the odors of the birds at the beginning and end of the study. Right away, we found some striking patterns. In particular, the birds in the antibiotic group produced much lower amounts of several volatile compounds after treatment. Using a statistical modeling method that could identify specific influences on the changes in volatile profiles, we found that the greatest change in the antibiotic treatment group was a reduction in three methyl ketones (2-tridecanone, 2-tetradecanone, and 2-pentadecanone). That decline appeared to be caused by the reduction of Staphylococcus and Pseudomonas, plus a few other bacteria.

If you're paying really close attention, you may have noticed that those three methyl ketones are the same ones that correlated with reproductive success in juncos. We had discovered that males tend to have higher concentrations of these compounds than females. And males who had more of these compounds (thus

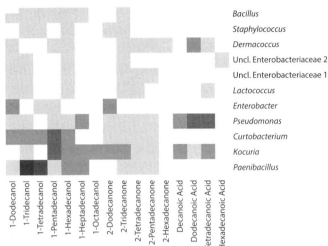

A heat map showing volatile compounds detected in bacteria cultivated from junco preen oil. *Adapted from* Whittaker et al. 2019, Journal of Experimental Biology 222:jeb202978

smelling even more "male-like") sired more offspring than males with lower levels. At the same time, females who had less of these compounds (a more "female-like" odor) had more offspring than the others. We had revealed that the compounds that seem to be important in the birds' reproductive behavior are actually produced not by the birds themselves but by symbiotic bacteria!

These results were very exciting because they were the first evidence that bacteria were indeed the source of junco chemical signals. However, I have to be honest—the results weren't quite as clear cut as they sound here. Because of small sample sizes and a high degree of variation among the birds in the study, statistically speaking, our results weren't super impressive. Ideally, in a study like this one, you want to show a clear interaction between time and treatment, where the subjects that received the experimental treatment (in this case, the antibiotic group) were significantly different after treatment, and the control group was basically the same before and after. Unfortunately, our small sample size meant that natural variation within the groups made it hard to detect such clear statistical patterns. The control group also showed changes before and after treatment for unknown reasons. Plus, due to random chance, on average the odors of the birds chosen for the antibiotic group were already different from the birds chosen for the control group. In a perfect scientific world, there would have been no difference between the groups at the beginning of the study, but due to time and funding constraints, we weren't able to test them before we began the experiment.

What was a certainty, however, was that antibiotics couldn't wipe out all of the bacteria that were present, and the birds did still produce odors, albeit changed ones, after treatment. How could we be absolutely sure that the bacteria were the primary source of these volatile compounds? We needed one more study— but this time we'd bypass the whole bird bit and measure the odors of the bacteria themselves.

CULTURED SMELLS

In the summer of 2017, I returned to Mountain Lake Biological Station once again to spend a week collecting preen oil from adult juncos. But this time, I didn't freeze the samples. Instead, I put the preen oil into special tubes designed for transporting samples with live bacteria. Kevin Theis and I were going to grow these bacteria and measure the odors they gave off.

At the end of the week, I carefully packed up the samples and drove eight hours straight from Pembroke, Virginia, to Wayne State University in Detroit, Michigan, where Kevin had landed a tenure-track faculty position in the medical school researching the human microbiome. Once there, I parked (illegally) near the building's entrance and called Jonathan Greenberg, a graduate student in Kevin's research group, who was awaiting my arrival. I handed off the samples into his gloved hands, and he immediately took them to the lab, where he spread the preen oil onto petri dishes, which he then quickly popped into an incubator.

I noted earlier that with standard laboratory methods only certain kinds of bacteria could be cultivated, making it nearly impossible to fully characterize a bacterial community using cultivation methods. Fortunately, methods have improved greatly in recent years, with the development of different kinds of growth medium (the nutrient-rich gel that the bacteria grow on) and more advanced ways to control the air around the plates. Because our samples came from outside the gland, where they would be exposed to air, as well as inside the gland, where there was no oxygen, Greenberg cultured our samples in both aerobic and anaerobic conditions. Using both of these methods would maximize our chances of growing bacteria that represented the natural uropygial gland communities.

Bacterial colonies grew successfully on several of our plates, mostly on those under aerobic conditions, but a couple of plates in the anaerobic conditions also showed growth. As in our other studies, Greenberg sequenced the 16S rRNA gene from these bacterial cultures to identify what we had grown. We successfully cultivated twelve types of bacteria, including *Pseudomonas* and *Staphylococcus*, which were the two that had the biggest effect in our antibiotic experiment. We also cultivated several other bacteria that we had seen in our previous surveys of natural uropygial communities, including Enterobacteriaceae, *Stenotrophomonas*, and *Lactococcus*.

The final goal was to learn what kind of volatile compounds the bacterial colonies could produce on their own, away from the juncos. We sent samples to our chemistry collaborators at Indiana University, Helena Soini and Milos Novotny. Soini used gas chromatography–mass spectrometry to measure volatile compounds given off by the bacterial cultures. Sure enough, these volatiles included all of the same compounds that we had been studying in junco preen oil. Every type of bacteria that we had cultivated produced at least a few of our junco volatiles of interest— completely independent of the juncos themselves. The biggest producers were *Pseudomonas, Bacillus, Curtobacterium, Kocuria,* and *Paenibacillus,* each of which produced large amounts of several volatile compounds typically associated with junco chemical communication.

We had finally answered my crucial question! We had discovered that symbiotic bacteria in junco uropygial glands produced the volatile compounds in preen oil that communicated information about junco identity and quality. But as always in science, this triumph just led to more mystery. Bacterial communities change constantly in response to changes in their environment and transmission from other communities. And uropygial gland bacteria are on the surface of the body, especially vulnerable to incoming bacteria from the environment and from social partners. How much did these communities change, and how were they still able to produce informative signals about their hosts?

I realized that in order to keep following these questions, it was time to reframe how I thought about animals and the bacteria that live on them. We have a habit of thinking of microorganisms as separate from ourselves, yet we know now that microbes are deeply involved in most if not all of our biological processes. Instead of thinking of hosts and their microbiomes as separate organisms with different agendas, we should consider how they work and evolve together as a single unit.

THE HOLOGENOME THEORY OF EVOLUTION

All animals are home to symbiotic microbial communities. However, in the past, microbes were seen only as harmful agents that caused disease. Most scientific research (as well as cleaning and personal hygiene products marketed to the public) has traditionally focused on pathogens and how to get rid of them. But in recent decades we have learned a stunning amount about the beneficial services that these microbial communities provide to their hosts—everything from assistance in digesting food, to protection from pathogenic microbes, to enabling our nervous systems to develop properly, to producing scents used in communication. It's becoming obvious that animals and their microbiomes have evolved these relationships together over millennia. Because hosts and their microbiomes are dependent on each other for survival and reproductive success, one theory suggests we should be considering these assemblages of organisms as a single entity: a holobiont (from the Greek hólos, "whole," and biont, an English suffix used in biology to refer to a living organism). Similarly, the hologenome is the collection of all of the genes present in the holobiont.

Populations evolve through patterns of variation, selection, and inheritance. In order for a trait to evolve, there must first be natural variation in the trait. Selection is the process by which some of these variants are more successful than others: successful variants survive and reproduce, while unsuccessful variants die before reproducing, or have fewer offspring. Finally, inheritance of the variant is required. If the trait is not genetically based, it cannot be passed on to the offspring, and it will not influence the success of future generations. Through these three processes, beneficial traits are able to increase in frequency in a population over time.

The hologenome theory of evolution suggests that since animals (or plants, or other living hosts) and their symbionts are reliant upon each other, we should consider them as a single unit in the evolutionary process. Hosts that provide favorable environments for their symbionts can nurture healthy populations of microbes, which in turn provide services that aid in the survival and reproduction of the host. The host's behavior can help spread these microbes to other hosts, assisting the survival and reproduction of the microbes. Thus, the host and the symbionts need to work together to survive together.

Many such relationships have already been discovered. Research on the gut microbiome, for instance, is referenced everywhere from television commercials hawking vegan smoothies to the cover of *Science* magazine (January 5, 2018, issue). Gut bacteria are necessary to digest plant matter. The tough cellular walls of plants cannot be broken down by animal enzymes and instead require fermentation by bacteria to chemically dismantle them into nutrients that we can absorb. This process feeds both the microbes and the animal hosts. Animals that lack the proper microbes are not able to access a substantial portion of the energy available in their environments, making transmission of these microbes from parent to offspring very important. In mammals, the fetus develops in sterile conditions inside the placenta, with no contact with the mother's microbiome.* Transmission begins at birth: some microbes are passed directly from mother to offspring as the newborn passes through the birth canal. Others may be acquired through consumption of the mother's milk, or transferred via physical contact.

Louis Pasteur is credited with the idea of experimentally raising animals completely lacking a microbiome, so-called *germ-free* animals. In 1885, he speculated that such animals would be unable to survive, given the evolutionary relationship between hosts and symbiotic microbes. After World War II and the invention of antibiotics, the idea of creating germ-free animals—and even people—appealed to the public's imagination, particularly in the realms of science fiction and space travel. Some were concerned about the possibility of introducing terrestrial microbes to extraterrestrial

---

* This finding is one example of a long-held belief being ultimately confirmed. Kevin Theis and his collaborators published a number of studies that found no placental microbiome in mice, rhesus macaques, or humans.

life forms, arguing that perhaps space exploration should only be conducted with germ-free humans. The converse—humans being infected with extraterrestrial pathogens—was the topic of the 1969 best-selling novel The Andromeda Strain by Michael Crichton.

Germ-free animals are not science fiction, however. In the late nineteenth century, scientists created the first germ-free guinea pigs by using a sterile caesarian section method, keeping the newborns from coming into contact with the mother's microbes. They were able to keep the newborns free from microbes for two weeks. Later, laboratory mice became the most common germ-free animals in research.

Studies with germ-free mice have revealed that the microbiome has a profound influence on many biological functions, including metabolism, digestion, immune response, and nervous system response. In 2004, a group of researchers at Kyushu University in Fukuoka, Japan, were the first to demonstrate the influence of microbes on brain function. They found that germ-free mice produced substantially higher levels of corticosterone when placed in a tube for one hour. This hormone, secreted through a process involving the hypothalamus, the pituitary gland, and the adrenal glands, has many important effects on the body, stimulating the fight-or-flight response and mobilizing energetic resources to allow an animal to quickly respond to a potentially life-threatening situation. In the short term, this response is necessary to ensure survival. However, because of the way corticosterone mobilizes the body's resources, having levels of this hormone elevated beyond what is needed in the moment becomes harmful to the animal. We humans are often warned of the dangers of chronic stress and the toll it takes on the body, including suppression of the immune system and increased risk of heart disease and depression. The finding of elevated stress in germ-free mice was groundbreaking, suggesting that a healthy microbial community could be very important in brain development and mental health.

Since then, more research has shown that the gut microbiome influences the brain and behavior, leading to the concept of the microbiome-gut-brain axis. Although we do not yet understand all of the mechanisms, it is clear from studies of germ-free animals that the microbiome influences stress responses, social behavior, and cognition. These differences may be caused by changes in the expression of genes that affect the activation of neurons in the brain.

Learning about the role of microbes in our own biology made me realize that I had been thinking about bacteria the wrong way my whole life. Like most people, I had spent most of my life thinking of them as gross, disease-causing creatures that we needed to eradicate. But, in fact, most bacteria are actually neutral, going about their own business and having little effect on us, while others are beneficial. Thanks to a prominent advertising campaign on television, many of us now know that eating yogurt is good for your digestive system because it contains helpful bacteria. But even once I understood that we had many positive relationships with our microbes, I still thought of them as separate animals, interacting with us the way that other people, pets, or pests do.

Our relationship with microbes is different than our relationships with other living beings. They are part of us, actively participating in our growth, nutrition, and immune function, helping shape who we become. We've evolved with them.

Being introduced to the idea of the holobiont made me reconsider not only how I thought about my own relationship with microorganisms but also how I approached research on chemical ecology in birds. I began to shift my focus. Instead of simply asking whether bacteria were responsible for the odors used in communication, I wanted to understand more about the relationship between the birds and their microbes. The big picture issues I'd had in the back of my mind for months came rushing back, this time informed by my new research findings. Now I was prepared to ask some seriously far-reaching questions: Where did the birds get their microbes? And how did birds' relationships with bacteria affect their behavior, and even their evolution?

Many of the bacteria that we found in uropygial glands were also commonly identified in soil samples. Juncos are ground nesters and ground feeders, spending quite a lot of their time in contact with the soil. It made sense that they would have a lot of soil microbes in their microbiomes. How did junco microbiomes compare to the microbiomes of other birds? How much of a young bird's microbiome was inherited from its parents and how much was just picked up from the external environment? Does an adult's microbiome change over time as a result of their behavior—such as whether the bird migrates, who it reproduces with, or whether it has to compete with others? The questions were piling up fast, and my ideas for new experiments were accumulating right along with them.

CHAPTER 6

THANKS FOR SHARING

The smell of a body is the body itself which we breathe in with our nose and mouth, which we suddenly possess as though it were its most secret substance and, to put the matter in a nutshell, its nature. The smell which is in me is the fusion of the body of the other person with my body; but it is the other person's body with the flesh removed, a vaporized body which has remained completely itself but which has become a volatile spirit.

—JEAN-PAUL SARTRE, *Baudelaire: Critical Study*

HANDSHAKES AND HOUSEMATES

Whenever you touch another person, some of your microbes rub off on them, and some of their microbes relocate to you. We share microbes not only with our close family and friends but also with people who we only encounter briefly. A simple handshake can transfer hundreds, if not thousands, of bacteria and viruses between people.

In some hospitals, handshake-free zones have been established to reduce the risk of infection in crucial areas like the neonatal intensive care unit. This policy can be very effective in reducing disease transmission in a healthcare setting, where pathogens may be more common than in typical public spaces. However, at least one study found that transmission of pathogens in highly social situations was actually quite rare. The researchers swabbed the hands of fourteen Maryland school officials before and after grade school and high school graduation ceremonies. The officials participated in several ceremonies each, with a grand total of 5,209 handshakes among them. Of all of the postceremony samples, only two contained pathogenic bacteria, one of which was methicillin-resistant Staphylococcus aureus (MRSA). These results could indicate that, in many contexts, perhaps there is less reason to be concerned about sharing germs with people than we tend to assume. Still, during flu season or a pandemic, avoiding handshakes will reduce your risk of picking up these rapidly spreading pathogens.

Most of the microbes we share are harmless, of course, and many are even beneficial. Just as we've seen in animals, some of these microbes contribute to a person's odor and convey chemical information about that individual to others. In fact, it turns out that the handshake may be the human equivalent to canine butt-sniffing.

You may not realize it, but you sniff your hands all the time. Frequent face-touching has been considered a so-called displacement behavior in response to social stress. Displacement was once a common explanation for actions that appeared to serve no function—like a "nervous tic." However, researchers found that as we touch our faces we are also more than doubling our nasal airflow: although it may not be conscious, we are sniffing. As it turns out, after shaking hands with another person, we touch our faces and smell our hands even more. In one study, participants were filmed alone in a room before and after an experimenter entered the room and introduced herself, shaking hands with the participants. The researchers counted how often and for how long the participants raised their hands to their faces and found that hand-sniffing increased significantly after shaking hands. Interestingly, there was a gender component to the results: when shaking hands with someone of the same gender, the participant spent more time sniffing their right hand, the one used in the handshake. When the participant shook hands with someone of a different gender, they actually increased the time spent sniffing their left hand, which was not used in the handshake. The researchers suggested that sniffing the right hand was for investigating the scent of the handshaker, while sniffing the left hand meant the participants were comparing their own scent to the other's.

Our skin microbiome is our primary interface with the world around us, and it reflects our interactions with our physical and social environments. Likewise, everything we encounter collects our microbiome, because we are constantly shedding skin cells and touching surfaces. Not surprisingly, humans who live together have very similar microbiomes. In a study of seven families and their homes over the course of six weeks, researchers found that people who lived together had similar skin microbiomes, and the bacteria on the surfaces of their houses matched the bacteria carried by the family. In a house with one couple and a third roommate, the couple had more similar bacteria to each other than to the roommate, but in homes with married couples and their children, all of the family members were almost equally similar. The part of the body that was most similar among housemates? The bottom of the foot.

Three of the families moved to a new home during the study. Once they had settled into the new home, the microbial communities on the surfaces matched those of the old home they had left—even in one case where the previous dwelling was a hotel room. When a household member left for as few as three days, that person's microbial signature on the house surfaces diminished greatly. Clearly, people spread their bacteria around their environments extremely quickly, and their microbiome takes over the previous microbiome.

In another study, researchers compared the skin (forehead and palms), mouth, and gut microbiomes of sixty families (including couples with children, dogs, both, or neither). In all of the microbial community types that were sampled, family members were more similar to each other than to people from other families, but it was their skin microbiomes that had the highest similarity. Humans also shared a significant portion of their skin microbial communities with their dogs, and owners were more similar to their own dogs than to other people's dogs. Interestingly, adult couples who were dog owners had more similar microbiomes than adult couples without dogs. Perhaps interacting with the dog led to sharing more microbes among them all? Adults with dogs also had more diverse skin microbiomes than adults without dogs. Much of this diversity was due to the presence of bacteria that is found primarily in dogs' mouths, as well as soil bacteria presumably tracked in by dog feet. Owning cats did not affect human skin microbiome diversity or similarity, although the cats themselves were not sampled for this study. Having children also did not affect adult microbiome similarity or diversity.

In everyday life, most people may not spend much time in physical contact with other people who are not family members or romantic partners. They likely limit the number of people with whom they share their microbiome. However, I can think of one notable exception: people who participate regularly in full contact sports.

FULL CONTACT

When I began my postdoc at Indiana University, I fell victim to the belief that academics have to dedicate every waking moment to their research if they want to be successful. I worked as hard as I could, and I spent my little remaining time worried that I wasn't working hard enough. After a few months of this high-stress pattern, I had my first (and so far, only) panic attack, leading to an emergency room visit because I thought I was having a stroke. Fortunately, I was fine, but this terrifying event was a serious wake up call. I knew it was time to make some changes. I needed to learn what work-life balance really meant. At the time, my primary extracurricular activities were drinking cocktails and watching movies on my couch with my spouse—not exactly a healthy antidote to all the work stress. I needed to find something else.

A graduate student in the IU Biology Department was on a *roller derby* team. I didn't know her personally, but I knew this fact about her because she sent emails to the Biology Department email list whenever her team had a game. I wasn't sure what roller derby was exactly, but in my head, I filed it away with stuff like burlesque clubs: activities that some women found empowering and tried to convince me that I should like, but that I personally found exploitative. Another member of the team worked in the same department as my husband, Nathan, and he suggested we go to a game sometime. I just shrugged.

But then, like thousands of other women in 2009, I watched the movie Whip It. Adapted from the novel *Derby Girl* by Shauna Cross, the movie tells the story of a disaffected high school girl, Bliss Cavendar (played by Elliot Page), who discovers roller derby in Austin, Texas. Bliss finds her home in the fast-paced and sometimes violent sport, buoyed by a league full of supportive and fierce women. Her life as a people pleaser with a chronic case of FOMO (fear of missing out) is transformed by derby, as she discovers her own strength and true sense of self. Suddenly, I found myself quite intrigued. I turned to Nathan on the couch as the credits rolled. "OK. Let's go to a game."

I sat in the stands at the Twin Lakes Recreation Center, beer in hand, transfixed as the skaters knocked the hell out of each other.

It looked so *satisfying*. Roller derby in its modern form is usually played on a flat oval track, around which two teams of skaters race. Each team consists of four "blockers" and one "jammer," whose goal is to score points by lapping the other team. The blockers work together to prevent the opposing team's jammer from passing, while also trying to help their own jammer break free from the pack. It's not an all-out brawl, despite the sport's vicious reputation: there are rules prohibiting moves like hitting an opponent in the back or smashing them with your head.

Staring at the track, entranced, I announced, "I think I want to do this." Next to me, Nathan smiled. "I know," he replied, irritatingly smug. But at thirty-five, I was pretty sure that I was older than most of the women on the track, and I had spent almost my entire adult life disdaining "exercise."* As a result, I was extremely out of shape. I just wasn't sure that derby was a realistic endeavor for me to attempt. And, I thought ruefully, a broken leg would probably put quite a crimp in my fieldwork plans.

But wait. In addition to the players, there were seven referees skating on the track, blowing whistles forcefully and sending players to the penalty box at will. *And they were all men.* As were the coaches. "Not very empowering, if none of the people calling the shots are women," I thought. An idea bloomed. I'd have to learn to skate first, though—an intimidating prospect.

When we moved to Michigan a few months later, the new job as managing director gave me an opportunity for a fresh start, and motivation to start living in a different way. Nathan and I joined a gym, choosing a YMCA with an indoor roller hockey rink. We only managed to make ourselves work out once a week at first, using the equipment in the upper-level studio. The upper level looked out over the first-floor rink, though, and one day I saw the

---

* I was quite fond of saying "I don't run, because I might trip and spill my martini." I thought I was very clever.

Lansing Derby Vixens skating on the track below us. They looked amazing and strong and fearless. I wanted to be one of them, but I didn't think I was ready—surely, I should learn how to skate first, or at least get into better shape. I ordered a "fresh meat" package of skates and protective gear online, but then I was too afraid to even try them on.

We bought tickets to the next Vixens game. I was surprised by the size of the crowd, which was much larger than the one I had been part of at the rec center in Bloomington. In 2011, when derby was new to Lansing, the Vixens could sell hundreds of tickets. During the second game we attended, I finally got brave enough (thanks to a few gin and tonics) to type out a message on my Blackberry and send it to the recruiting email address on the program. Before long, I got a response with information about how to become part of the team.

That winter, I joined the Derby 101 bootcamp, where skaters from the team taught newbies how to play. It was *hard*. I was even more out of shape than I had realized. At my very first two-hour session, I fell down so much and so hard that I am fairly certain I broke my tailbone. When I got home, I was so sore that I had to crawl up the stairs on my hands and knees. I sat in a hot bath and cried for at least an hour. Nathan said gently, "It's OK if you don't want to go back next week." I glared at him and snarled, "Oh, I'm *fucking going back.*" I was hooked.

Of the eighteen beginners in my class, I was the slowest to learn to skate. It was months before I wasn't a danger to others with all my flailing. But I kept at it. Derby quickly became a huge part of my life. Now, a decade later, not only have I learned to skate, I have officiated well over six hundred games in five countries on three continents. I have even been a tournament head referee for eight multiday events, including one of the Women's Flat Track Derby Association International Playoffs in 2019.

Women's Flat Track Derby Association certified referee Chunk Rock Girl (that's me!) officiating at the 2017 Besterns tournament in Denver. *Photo by Joel Giltner*

I've also continued to pursue the goal of overcoming the male bias in officiating. I work to recruit and mentor new women and nonbinary referees, staffing them in tournaments and giving them opportunities to work at higher levels. And I've been openly denouncing sexist and harassing behavior by other officials (who were primarily men), which is one of the biggest reasons that officials from underrepresented groups leave the sport.

It's become a cliché in my circles to say "roller derby saved my life." But it's often true. In my case, I had never been athletic, and I was very overweight. I didn't have many friends or activities outside academia, and I had lost touch with what the rest of the world was like. Now, I have a huge global network of derby friends from all walks of life. I lost over forty pounds and kept most of it off. I can bench press over a hundred pounds. I even took up running.*

---

* Even marathons. I do trip sometimes, but consolation martinis are waiting for me at home.

Roller derby didn't cure my imposter syndrome, but it definitely lessened it. After all, if I have the strength and endurance to skate for hours and keep getting up every time I fall, I feel like I can handle just about anything. And, as a bonus, I'm pretty sure participating in roller derby increased the diversity of my microbiome.

Roller derby is a full contact sport. Although you're not allowed to use your hands to block your opponents, you can use them to grasp your teammates. Participants are often pushed right up against each other for much of the game. Dr. Jessica Green, microbiologist and director of the Biology and Built Environment Center at the University of Oregon, was a former skater for Emerald City Roller Derby (ECRD), the local Eugene, Oregon, team. She was interested in how skin microbes are transferred by physical contact and decided that roller derby was the perfect environment to study this topic. In 2012, at the annual Big O roller derby tournament hosted by ECRD, Green and her lab members tested the players' skin microbiomes to track the effect of playing roller derby, sampling the upper arms of skaters before and after two games. ECRD played in both games, first against DC Rollergirls from Washington, DC, and then against California's Silicon Valley Roller Girls.

Before their games, teammates had very similar skin microbiomes, and the teams' microbiomes were significantly different from each other. I know from experience that most players rarely wash their required protective gear—knee pads, elbow pads, wrist guards, and helmet—which probably helps cultivate skin bacteria. The samples taken from the home team, Emerald City, were also similar to microbial samples taken from the surface of the track. What with all that crashing to the ground, it's not surprising that the home team would share bacteria with their track.

But over the course of a one-hour game, the skin microbiome of all of the players changed. By the end of each matchup, the microbial samples of players from different teams were much more similar to each other. They had shared more than the thrill of the clash.

The researchers didn't follow up with the players after the games, so we don't know how long these changes to their microbiome lasted. Did they revert to their typical skin microbiome after a shower? Or do some of the new arrivals persist, so that the skin microbiome retains a record of one's past activities? It's interesting to think about how these short-term encounters might affect not just a person's microbiome but also the chemical signals produced by the microbes that make it up. Does our microbiome and, therefore, our scent simply reflect our social behavior, or does it contain fundamental information about who we are?

ALL IN THE FAMILY

Around the same time that Jessica Green and her group were sampling roller derby teams, I was also collecting data on how microbes are shared by physical contact. I was inspired by Kevin Theis's finding that hyenas living in the same group all had similar microbes in their scent pouches as a result of scent-marking the same surfaces. Nestling birds spend the first several days or weeks of life huddled together in a tiny nest, and I expected that they would all have similar bacteria as a result. I also expected that those bacteria would, at least initially, come from their parents. But about a quarter of all junco nestlings are not related to the male that is raising them. Which was more important in determining microbiome similarity: genetic relatedness or physical contact?

In the summer of 2012, I sampled nine nests at Mountain Lake Biological Station, taking DNA and microbiome swabs from the parents and all of their nestlings. Using paternity tests, I found that only three of the nests had no extra-pair young: in those cases, the mother and father taking care of the nestlings were also the genetic parents all of the nestlings. Two of the nests had only extra-pair young, so that the male attending the nest was not related to any of the offspring he was raising. However, within each of those nests, the nestlings all had the same genetic father,

: 153 :

so they were full siblings. Finally, the remaining four nests had a mix of within-pair and extra-pair young, so the males were related to some but not all of the nestlings they were raising, and the nests contained some full siblings and some half siblings.*

Family members all had very similar bacterial communities in their uropygial glands, whether they were related or not. Adults and nestlings were somewhat different from each other, probably because nestlings were confined to their nests, while the adults encountered a broader environment. Given that the nestlings were all squeezed together in the nest, twenty-four hours a day for twelve days, we would expect them to share bacteria. And if those bacteria were indeed the source of their preen oil odor, we would also expect them to smell the same, regardless of whether they were half siblings or full siblings. So we also analyzed their preen oil chemical profiles, and the same pattern held: members of the same family groups, whether related or not, had similar odors.

I had expected to find sex differences, at least in the adults, given that my previous studies had shown sex differences in the odor of male and female juncos during the breeding season. But it was during this project that I discovered that sex differences in odor did not persist throughout the entire breeding season. There was no difference between the bacteria or the preen oil odor of males and females. Instead, the mated pairs were very similar to each other. Some steroid hormone levels, such as testosterone, decrease while birds are raising offspring, since they are not yet ready to mate again. Perhaps the lack of hormonal differences between the males and females while caring for nestlings contributes to them smelling alike. Or perhaps they share bacteria while working together to raise a family, and that also results in smelling alike.

---

* In all cases, the mother raising the nestlings was also their genetic mother. Although egg-dumping—laying your eggs in another female's nest—occurs in some species of birds, it is not known to occur in juncos.

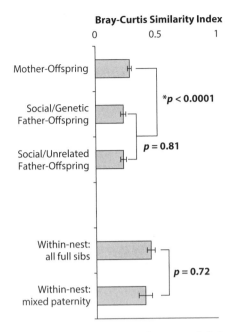

Average Bray-Curtis Similarity Index measure for uropygial gland bacterial communities in junco nests. Bray-Curtis Similarity is a measure from 0 to 1, with higher numbers indicating greater similarity. The most similar family members were siblings, whether they shared two genetic parents or just one. Mothers were significantly more similar to their offspring than fathers. Genetic relatedness did not affect how similar a father's bacterial community was to his offspring. The probability that the compared groups are statistically the same is represented by p; an asterisk indicates that they are significantly different (p less than 0.05).

When I compared within-family similarity in bacterial communities, siblings had the highest levels of similarity to each other, whether they were full or half siblings. Mothers, who spent significant amounts of time sitting on the nest keeping the babies warm, were more similar to their nestlings than fathers were. Only the female incubates eggs and broods nestlings in many bird species, including juncos, so they have much more contact than the fathers. The fathers do assist in feeding the nestlings, so they have some contact with them via those interactions. Fathers, whether genetically related or not, also had bacterial communities similar to those of their offspring. Most notably, being the genetic sire of the nestlings did not increase the males' microbial similarity to them.

In most birds, the uropygial gland isn't fully formed when the birds first hatch. It develops over their first few days outside the egg. I tried to sample the nestlings multiple times throughout the nesting period so I could understand more about the development of the microbial communities in their glands. I managed to collect preen oil samples from six-day-old juncos, when they were about halfway through the total time they spend in the nest growing and developing. Taking samples from three-day-old chicks was significantly more difficult because many of the chicks were not yet producing much, if any, preen oil. So, it was clear that birds do not hatch with an intact uropygial gland bacterial community that somehow got started while they were in the egg. Instead, they have to pick it up from the environment around them. It seems that an organism's bacterial community is shaped by social behavior, rather than one's own genome.

This same result was found in a study of Eurasian hoopoes published before ours, though after we had collected our data. Mothers and their nestlings had similar bacterial communities, although there were some age-related differences between those groups. Nestlings reared in the same nest had similar bacteria, even when they had been cross-fostered—that is, taken out of their original nest and placed with a different bird family. Nestlings do not appear to inherit their uropygial bacterial communities through vertical transmission, meaning inherited directly from the mother. If bacterial communities were vertically transmitted, we might expect that some of the mother's bacteria would be present in the eggs, and nestlings would hatch already having bacteria similar to their mothers. Instead, new chicks pick up their bacteria through horizontal transmission—from the physical and social environment around them.

Not all birds show this same pattern of microbiome similarity with physical environment or within groups. Leach's storm petrels, a type of small seabird that breeds in the northern Atlantic and Pacific oceans, dig underground burrows to build their nests in. Oddly, there was no relationship between an adult petrel's uropygial or skin microbiome and the microbiome of its underground burrow, even though they must come into contact with the soil often. And the birds' microbiomes were not more like their mate's than a randomly chosen bird from the population. Instead, a bird's sex was the primary driver of its microbiome compositions, with females having higher overall microbial diversity and males having higher abundance of certain bacteria. Why are these birds so different from the juncos and hoopoes? Unlike songbirds, who share parental care and interact frequently while raising their offspring, petrels and other seabirds take turns caring for offspring, with one parent staying in the burrow incubating the eggs while the other is off foraging at sea for days at a time. They have very little physical contact with each other, and so they probably aren't sharing microbes as intensely as the songbirds do.

Because my main interest is in chemical communication, I've focused mostly on the uropygial microbiome and the skin microbiome. But many studies have found that groups of animals also share gut bacteria. Unlike mammals, which have separate digestive and reproductive tracts, birds, amphibians, and reptiles have a cloaca, which is used for both excretion and reproduction. Bacteria from the gut can be found in the cloaca, as well as sexually transmitted bacteria. In juncos, cloacal bacterial communities showed the same patterns of similarity within groups, regardless of genetic relatedness. We also found that nestlings' developing uropygial communities appeared to be a subset of the bacteria found in their cloaca. In newly hatched chicks, the cloaca and uropygial gland are right next to each other, and they move apart as the chick grows. Bacteria from the cloaca could likely colonize the uropygial gland in young chicks easily, which would give them their "starting point" for their uropygial communities. From there, the bacteria that are suited to living in the uropygial environment would thrive, and any bacteria acquired from the environment would have to compete with those starting bacteria. Such developmental processes can result in bacteria being shared across different microbial communities within the same body, with interesting implications for how sharing bacteria on the body surface could affect internal bacteria.

SEX, GROOMING, AND GUT BACTERIA

Microbiomes associated with specific body parts—the skin microbiome or the eye microbiome, for example—are not completely independent of each other. What's on the outside of the body can end up on the inside, and what's inside the body may later be found on the outside. Bacteria in the gut, for instance, can influence the skin microbiome and vice versa. This exchange means our social partners can affect not only our surface microbiomes but also our internal bacterial communities.

A pair of researchers, Subhash Kulkarni and Philipp Heeb, then at the University of Lausanne, tested how quickly bacteria on a bird's feathers could move to the digestive tract, and whether the bacteria would be transmitted to another bird. The researchers applied a nonpathogenic soil bacterium, Bacillus licheniformis, to the neck and wing feathers of zebra finches. They then housed each bird in a cage with a zebra finch of the opposite sex and sampled the bacterial communities on each bird's feathers and cloaca over time. The researchers found that the Bacillus bacteria appeared in the cloacae of the bacteria-painted birds within twenty-four hours, most likely because they ingested the bacteria while preening. The birds' cage mates also had the bacteria on their feathers and in their cloacae within just twenty-four hours. The bacteria may have been transmitted by allopreening, a behavior in which the birds preen each other, or by sexual behavior. Clearly, social and sexual behavior can spread bacteria in a very short period of time.

Duke University scientist Dr. Jenny Tung is also interested in how social behavior spreads bacteria through groups of animals. As part of a long-term study on yellow baboons in Amboseli National Park, Kenya, Tung examined the gut microbiomes of forty-eight baboons across two social groups. Over the course of one month, Tung's team of researchers collected fecal samples and recorded the diet and social behavior of the animals. The strength of social connections was measured by how often the baboons interacted. Baboons that spent the most time grooming each other had the strongest social connections. Baboons in the same social groups had very similar bacterial communities, and within the groups, those that groomed each other more frequently had more similar gut bacteria. These social relationships were the strongest influence on bacterial communities: neither age, nor sex, nor even dietary differences appeared to make a difference.

In previous studies that found humans who lived together had similar gut microbiomes, the researchers had attributed the similarity to the fact that people living in the same house tended to eat the same foods, but they had not specifically tested the role of diet. Tung's study of baboons was the first to take diet into account while simultaneously measuring all social interactions. The two social groups in the study had largely similar diets, since they lived in the same habitat. In both groups, about 85% to 90% of the diet consisted of grass corms (the underground part of a blade of grass), grass seed heads, acacia pods, and leaves. The primary difference was that one group consumed more fruit than the other (about 8% of total diet compared to only 2%). But none of the differences in gut bacteria were what you would expect if they had been related to increased fruit consumption. For example, there was no increase in bacterial enzymes commonly associated with digesting these foods.

Another group of researchers examined the gut microbiomes of a group of chimpanzees in Gombe National Park, Tanzania, over a period of eight years. Chimpanzees have what is known as a fission-fusion social organization, in which members of a large social group form smaller subgroups at different times, instead of staying together all day, every day. At Gombe, the chimpanzees spent more time together in larger groups during the rainy season (November through April). In contrast, during the dry season (May through October), food was less abundant, and they were much more often foraging alone or in small groups. The chimpanzees all had more similar gut microbiomes during the rainy season, when they were more sociable, than during the dry season, when they spent more time alone. Although there were different foods available in the seasons, the types of bacteria found in the fecal samples did not consistently differ between the rainy season and the dry season, indicating that social behavior, not diet, appeared to be the primary factor in shifting gut microbiome composition. Over the course of the study, several of the chimpanzees reproduced, giving the researchers the opportunity to compare mother and offspring gut bacteria. The offspring were not any more similar to their mothers than they were to other members of their groups. Again, the less obvious factor of social interaction was found to be the primary form of transmission. Maybe it's time to retire the aphorism "you are what you eat." It's not what you eat, it's who you know that matters.

FRIENDS WITH BENEFITS

The spread of microbes through social behavior appears to be a universal phenomenon. Dr. Michael Lombardo, a researcher at Grand Valley State University in Michigan, has gone so far as to suggest that sexual and social behavior may have evolved in animals *because* it increased the transmission of beneficial bacteria. In Lombardo's evolutionary scenario, animals who had a tendency to be social would have an advantage over those who were more solitary, because their behavior would allow them to acquire and share beneficial microbes. Having the right microbes would improve the host animal's biological functions, including digestion of otherwise inaccessible nutrients, proper development of the nervous system, and stronger immune function. These sociable animals would survive longer or reproduce more efficiently than their solitary counterparts, and over time, the tendency to be social would increase via natural selection, spreading throughout the species.

Lombardo's idea first appeared as an explanation for why female birds have more copulations than would be necessary to fertilize their eggs. Although most songbirds are socially monogamous, they are often not genetically monogamous, and many birds will copulate with multiple partners. Males copulate with as many females as they can, but this behavior is rarely questioned. The benefits to males are clear: although copulating with more females does take more energy, it's typically not so much energy that the males' other functions are negatively affected. Through these exploits, males have the potential to produce many more offspring than if they stayed faithful to their social mate. Their genes, including any genes that made them more likely to engage in extra-pair copulation, would be more common in the next generation, and the tendency would evolve. But for female birds, the evolutionary benefits of copulating with multiple males seem less obvious, at least at first glance. Females are limited in the number of eggs their body can produce in a season, and in species like juncos with maternal care, they are further limited by how many offspring they can take care of—they can't raise multiple nests at once, for example. No matter how many males they mate with, promiscuous females won't be able to produce any more offspring than monogamous females. In fact, there may be disadvantages, such as transmission of disease or abandonment by one's jealous social mate.

Yet female birds often do seek out copulations with males other than their social partner, despite the additional energy required and risks incurred. Why? Several researchers have suggested that females may mate with extra-pair males to ensure their offspring will have higher-quality genes, especially if the female's social mate is low quality or genetically related to her. These hypotheses are attractive, but empirical support is mixed.

Suspecting that something else might be driving extra-pair behavior in females, my fellow Ketterson lab member Nicole Gerlach analyzed seventeen years of data from the dark-eyed junco study at Mountain Lake to look for benefits of extra-pair fertilizations. Gerlach found that extra-pair offspring had higher lifetime reproductive success than within-pair offspring. That is, a junco whose genetic sire was not its social father had more offspring of its own throughout its lifetime. So, the answer appears in future generations. Females who seek out extra-pair copulations do ultimately have higher reproductive success—by having more grandchildren!

Improving the quality or number of your offspring is considered an indirect genetic benefit. Lombardo, however, suggested that birds may also receive a more direct benefit to their own health from copulating with more partners and sharing beneficial bacteria. When animals mate, they also share the bacteria present in their reproductive tracts. As mentioned before, in birds, the reproductive tract ends in the cloaca, which is also the opening for the digestive tract, so mates may also share gut bacteria. Lombardo posited that engaging in frequent copulations with multiple partners may have evolved as a way to acquire and transmit beneficial bacteria. If these bacteria helped the host birds to survive longer and reproduce more, then the tendency to mate at these higher rates would be passed on to offspring and would increase in frequency in the population over time.

Nearly a decade later, Lombardo made the even grander suggestion that the evolution of sociality itself could have been spurred by the fact that it increases access to beneficial microbes. Explorations into why some animals live in groups and others do not have historically generated ecological explanations. For example, grass- and leaf-eating animals can often live peacefully in large groups because their main food resource is abundant. Such abundance means they do not have to worry about competing with group members over food. With that pressure removed, they can enjoy the group-living benefits of increased vigilance and protection from predators, improved ability to find food and mates, and assistance in raising offspring. Yet fruit-eating monkeys and apes, and carnivores like hyenas and lions, also live in groups, even though their primary food resources are more difficult to obtain, which often results in intragroup competition.

Primatologist Robin Dunbar hypothesized that the reason that monkeys and apes have such large brains relative to body size is not a result of any special mental demands required by their ecological circumstances but is instead due to the challenges posed by living in large social groups. Long-lived animals with consistent group membership need social skills to manage relationships and ensure that they all won't kill each other, even though group living may increase competition for some resources. To navigate complexities that arise, these animals have evolved solutions, such as the ability to recognize each other and remember past interactions, dominance hierarchies that resolve conflicts without violence, and cooperative behavior to reinforce relationships. In effect, Dunbar argued that large groups and large brains evolved together.

Microbe sharing as a driving evolutionary force for group living adds another interesting layer. In the 1980s, Katherine Troyer, who studied the digestive biology of iguanas, argued that fiber-fermenting gut bacteria are so critical for the survival of herbivorous animals that the need to transmit them to younger generations was the selective force that drove the evolution of social groups. Lombardo took this idea further, pointing out that bacteria benefit animals not only through improved digestion but also through increased protection from pathogens, making them beneficial for all animals, regardless of the animal's primary diet.

An interesting way to test this idea is to look at species that show variation in social behavior—sometimes they are social, sometimes they are not—and compare their success in the different situations. For example, female four-toed salamanders build nests and take care of the developing embryos. They can choose to build solitary nests and take care of only their own offspring, or they can lay their eggs in a communal nest, as 50% to 70% of all females do. Females who choose to take advantage of a communal nest can either leave the parental care to another female or share the duties with others. The embryos of salamanders and other amphibians are particularly sensitive to fungal infections, which can kill them, so providing protection from such infections is important to survival. Some female four-toed salamanders harbor bacteria on their skin that produce antifungal compounds, which can protect the embryos from fungal infections. In a study comparing solitary nests to communal nests, a far greater percentage of the communal nests (73%) than solitary nests (only 26%) included an attending female with skin bacteria that produced substances harmful to pathogenic *Mariannaea* fungi. Females who attended the nest provided care that included moistening and cleaning the embryos while weaving among them, likely transmitting bacteria or antifungal secretions in the process. Communal nests also had a lower rate of complete failure than solitary nests.

This idea that sociality promotes transmission of beneficial bacteria has also been tested experimentally. Honeybees and bumblebees live in large colonies, and the adults have a very specialized gut bacteria community. Most bee species are actually solitary, and they lack most of the bacteria found in the social bees' guts. Researchers hypothesized that these gut bacteria were important in protecting the bees from a common gut parasite, *Crithidia bombi*, which can have devastating consequences on queens who are attempting to start colonies. Infected queens are less likely to successfully produce a colony, and they produce far fewer offspring than uninfected queens. To test whether socially transmitted bacteria protect bees from this parasite, Swiss researchers Hauke Koch and Paul Schmid-Hempel raised bees from their pupal stage in isolation. They fed one group the feces from their home-nest mates, which is part of normal bumblebee development, thereby allowing them to establish a normal gut microbial community. The other group did not receive these bacteria. The bees from the group fed fecal material had a significantly lower rate of infection from the parasite C. bombi compared to the bees raised without normal gut bacteria.

Clearly, behaviors that transmit beneficial bacteria among groupmates can be advantageous to the animal's health and survival. In species with complex social systems, including primates and carnivores, it has been suggested that certain social behaviors like grooming, food sharing, and kissing could have evolved because they transmit microbes. Other researchers have suggested that altruistic behavior—in which one individual helps another, incurring a cost to themselves with no apparent benefit—may have evolved as a way to transmit symbiotic bacteria. This fascinating hypothesis has a downright stunning implication—that it is not the animals themselves but the *bacteria* driving the evolution of their behavior.

AUTOBIOGRAPHY OF A BACTERIUM

So far, we've primarily considered host-microbiome relationships from the perspective of benefits provided to the host. But bacteria are not merely accessories to larger animals. They also reproduce, compete for resources, thrive, survive, or die. Different scenarios may be beneficial or harmful to their success, as well. If microbes can manipulate the host's behavior, it would be advantageous to their genetic line to influence that host in ways that increase the survival or reproduction of other potential hosts.

Zebrafish (*Danio rerio*) are small, easily maintained aquarium fish that have become a model organism in the fish world. A group of researchers at University of Oregon and their collaborators were interested in how the host's biology—specifically, its immune system—affected transmission of gut bacteria between individuals. A benefit of studying model organisms in the laboratory, rather than groups of wild animals, is that you can control most aspects of their environment and focus on the variables of interest. In this case, the researchers were able to use genetic engineering techniques to knock out a gene involved in the immune system of the fish. The gene, called myeloid differentiation primary response gene 88, or MyD88 for short, is involved in activating immune responses to pathogenic microbes. These responses include inflammation and the production of antimicrobial substances. Comparing the "knockout" fish, who would be unable to fight against pathogens, with "wild-type" fish, with unaltered genomes, would allow the researchers to determine what effect the immune response had on transmitting gut bacteria among fish.

The researchers raised zebrafish of both types either in groups or alone. There were two categories of fish groups: either all the same type (wild-type or knockout) or both types of fish mixed. They expected that the group-housed fish would have more diverse microbiomes than the solitary fish, since they would be exposed to group transmission. Additionally, perhaps the knockout fish would have different microbiomes than the wild-type fish, since they couldn't defend themselves against pathogenic bacteria.

Their predictions were partly confirmed: the solitary knockout fish had different microbiomes from the solitary wild-type fish. And the group-housed fish generally had different microbiomes than the solitary fish. However, there were no differences between knockout fish and wild-type fish when they were housed in groups— regardless of what types of fish were in those groups.

Instead, the primary feature of the microbiomes in those fish was the increased ability of the bacteria to disperse between hosts. Just the fact that the fish lived in groups presented a huge advantage to bacteria that could move easily from fish to fish, expanding their populations and changing the structure of all of the microbiomes in the group.

What made certain bacteria better at dispersal? The researchers analyzed the bacterial genes present in the overall microbiomes to see which ones had become more common. In particular, there was a notable increase in genes related to bacterial movement, including genes related to chemotaxis (the tendency of a microbe to move toward or away from specific chemicals in its environment) or to the production of flagella (the whiplike structures that help bacteria swim).

The bacteria in the group-housed fish microbiomes were also more likely to have genes related to quorum sensing. Quorum sensing is a way that bacteria detect how many of each other are present in the environment, enabling them to turn certain genes on or off depending on their population density. In order to detect each other, the bacteria send out chemical signaling molecules at a low rate. When there are a large number of fellow bacteria, these molecules build up to a strong concentration in the environment, empowering them to trigger changes inside the bacteria. One well-studied type of quorum sensing is biofilm formation: if enough bacteria are present, they produce a gooey substance to stick themselves to each other, allowing them to coat a solid or liquid surface. The plaque on your teeth is an example of a biofilm. Biofilms can also coat nonliving surfaces, and they can be a major cause of infection in hospital settings when they form on materials like catheters or other tubes used in the body.

The zebrafish experiment demonstrated that a host's social be-havior influences what type of bacteria will be successful within it. Reading this study, I suddenly realized that in my own mi-crobiome work I had not been thinking about what the bacteria themselves were doing at all. Even though I had acknowledged the importance of bacteria in animal behavior, and even though I recognized the value of the holobiont theory, I still had only been thinking about the microbiome almost completely from the per-spective of the host animal, with the animal's behavior and needs as the focus. I suppose I had just imagined that whatever bacteria happened to be present would be along for the ride, and that bac-teria could serve as an indicator of what the host animal had been doing. But bacteria have behavior too.

At that time, I was still pursuing the question of how much of one's chemical signals reflected one's own biology, and how much was a record of one's social interactions via the bacteria one ac-quired. I designed a study that I thought would be able to tease apart these differences. It didn't quite turn out as I expected.

In 2017, Dr. Joel Slade came to Michigan State to work with me as a postdoctoral researcher. As part of his research for his PhD at Western University in Ontario, he had studied the relationships among MHC genes, preen oil odor, and mate choice in song spar-rows. He was interested in doing more work with preen oil and odor, and in expanding his research to include microbiome meth-ods. Although I didn't have a formal lab, I was able to secure fund-ing through BEACON to hire him as a postdoc at MSU for two years. I was nervous about whether I could provide him with a useful postdoctoral experience, given my atypical position, but he has since become a tenure-track faculty member at California State University Fresno, so I'd say it worked out well.

At Mountain Lake Biological Station, Joel and I conducted a study to research how social behavior and genetic background inter-act to affect microbiomes and odors. We captured twenty-four juncos from two different subspecies: half were from the Carolina subspecies that we typically studied every summer at Mountain Lake, and the other half were from the northern subspecies that wintered on the mountain but migrated back north for the breeding season. Using two different subspecies ensured that the birds would have different genetic backgrounds. Then we housed all of the birds in divided sections of the outdoor aviary, changing their housing configuration every eight days to study how microbes were shared in the social groups. First, we assigned them to four compartments according to their sex and subspecies: female Carolinas, male Carolinas, female northerns, and male northerns. After eight days in these groups, we then mixed them up into male-female pairs, with each pair getting their own compartment. Finally, we moved them back into the four flock compartments. We sampled their microbiomes and their preen oil every four days throughout this experiment. We expected that their microbiomes and their odor would change depending on which other birds they were housed with, and we were interested to see how much of their microbiome's social history was retained over time.

I don't have final results to share with you here, but in some ways, the preliminary results do support some of our ideas—birds that were housed together did become more similar to each other over time. However, we are finding other patterns that are revealing that the real story is more complicated than we realized.

All of the birds' microbiome communities shifted over the course of the experiment. Most notably, diversity decreased, and their microbiomes became dominated by a handful of bacterial species, including *Pseudomonas*, *Lactobacillus*, *Staphylococcus*, and *Campylobacter*. This happened in all of the birds, regardless of subspecies or sex. It seemed that perhaps just holding the birds in captivity affected their microbiomes.

Although we didn't perform the same kind of analyses as the group who studied zebrafish, perhaps we were also selecting for

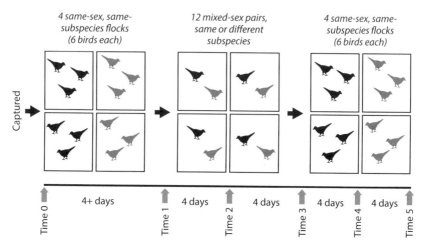

Schematic showing housing configurations for study to test effects of social history on Junco uropygial gland microbiome and preen oil odors. The timeline at the bottom indicates when we took preen oil and microbial samples from each bird.

dispersal capabilities in our junco microbiomes. *Pseudomonas* in particular made up a large proportion of the samples. These species are highly mobile, with a flagellum that helps them move quickly. In fact, some species of *Pseudomonas* are pathogenic and highly infectious.

Joel and I are still working through the data from this experiment, trying to understand the changes we observed. But these initial results made me wonder whether I was really studying what I thought I was studying—a question that all scientists should regularly ask themselves. Juncos in a cage, fish in a tank, and skaters on a roller derby track do share bacteria. I had been thinking of these scenarios as examples of how scent might reflect social history, but I needed to remember that microbes, especially those that cause disease, could be opportunistic.

PANDEMIC PERFUME

Everything in this chapter, with its emphasis on how social behavior leads to the transmission of microbes, feels especially relevant at this moment. I wrote the first draft of this chapter while self-isolated at home during the COVID-19 pandemic in the summer of 2020. The SARS-CoV-2 virus spread rapidly all over the world, and the only way to slow it down was to limit contact among people. "Social distancing" is now a common phrase, as people were directed to stay at least six feet apart from anyone they didn't live with and to avoid large gatherings. Schools and universities shut down, and those of us who could were urged to work from home. For over a year, most of my work and social interactions were via teleconference.

It was a frightening and confusing time, but these changes were necessary to slow the spread and "flatten the curve." This coronavirus is extremely transmissible. Although its death rate is unacceptably high from a human perspective, from the virus's perspective, it is quite low, allowing it to spread quickly through a population without killing too many of its hosts. A pathogen that kills its hosts too quickly does itself a disservice. If a host dies before the virus can move to another host, the virus dies with them. One of the things that made SARS-CoV-2 so successful is the fact that a host can be symptom-free for weeks, all the while transmitting the virus to others.

Unlike bacteria, viruses are obligate intracellular parasites. They are subcellular, meaning that they are simpler than a cell, and are made up of either DNA or RNA (COVID-19 is caused by an RNA virus) wrapped in a protein coat, which may have an envelope or protruding spikelike structures around it. They lack the rest of the cellular machinery needed to replicate themselves, and so they must infect the cells of a host in order to do so. Once they have entered a cell, they can commandeer the cell's replication machinery to replicate their own DNA/RNA, rather than the cell's own DNA, until the cell becomes full of copies of the virus. Depending on the virus, the virus can then bud off the original cell to go infect more cells, or the host cell itself bursts, releasing the viruses into the blood stream, where they can then travel to more host cells. This process can release hundreds, to thousands, to hundreds of thousands of viruses, all grown in a single cell.

Because viruses cannot replicate themselves, they cannot be cultured in a petri dish as easily as bacteria can. Methods of replication using cell cultures or embryonic chicken eggs have been developed for studying viruses. However, these methods are not ideal for diagnosing a viral infection, as they can take a very long time. To test for the presence of a virus, detection of antibodies in the host animal is typically used. Because viruses are subcellular, antibiotics don't kill them. Instead, we need vaccines, which stimulate our immune system to produce the correct antibodies that will attack the virus if we are infected.

We now know that COVID-19 is primarily airborne, with most transmission occurring through breathing in someone else's respiratory droplets or aerosols. Unlike the measles virus, which can continue to be infectious in the air for two hours after an infected person has left the room, the COVID-19 virus appears to be spread during immediate social contact, especially while indoors.

By limiting social contact to prevent transmission of this virus, we also shared less of our other microbes. A silver lining to all of this isolation is that I didn't get sick with anything for over a year. Of course, this lack of physical contact—and bacteria sharing—could also change the way we smell. On the online news site Vice, a 2020 article reported on the phenomenon of people noticing that their body odor changed during the quarantine. Because they were no longer physically interacting with coworkers, friends, or strangers, the composition of their skin microbiomes likely changed significantly, which translates into smelling different. Most people in the article reported smelling better, but one woman noted that her odor had gotten worse, because she smelled more like her cohabitating boyfriend.

But not to worry—if you find that you like your odor less as a result of social distancing, it probably won't last forever. North Carolina State University biologist Rob Dunn, who was interviewed for the article, assured us that as we return to interacting with society, our microbiomes will likely return to normal. Our bacteria ensure that we always carry within us the capacity to, as he put it, "once again become part of a bigger community of stink."

CHAPTER 7

MHC: MAGICAL
HAPPINESS
CONTROLLER?

*While many scientists would argue that a popular-level book
like this one should stick to established decades-old ideas,
my view is that nothing can be more exciting than what's
happening at the edge of knowledge.*
—DANIEL M. DAVIS, *The Compatibility Gene*

WHY DOES MHC MATTER?

My interest in the major histocompatibility complex, or MHC,
first arose when I learned of its possible ability to influence mate
preferences. The idea of a set of genes quietly controlling how an
animal, or a human, perceives the world and makes decisions is re-
markable. Unlike in physics, there are no immutable laws in biology. It's simply too complex for a single equation to explain all
instances of a particular phenomenon. To me, the idea of MHC's
role in behavior presented a seductive opportunity to grasp at a
unifying explanation of social and reproductive biology. But I

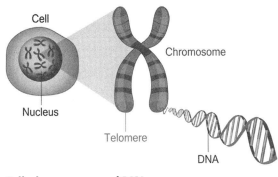

Cell, chromosome, and DNA.
Fancy Tapis / Shutterstock

soon learned that MHC is far more complex than I had initially appreciated.

Although the mysteries of MHC are a fundamental part of my work, and of the story of avian olfaction, until now we didn't need to delve too much into how it actually functions in the body. But to understand how MHC works in a more holistic way, we need to discuss some fundamentals of genetics. DNA is coiled into structures called chromosomes. Humans have twenty-three pairs of chromosomes. They come in pairs because you get one copy from each parent: a human egg has twenty-three chromosomes and a human sperm has twenty-three. Each chromosome carries a set of genes. A gene, as one of my graduate school professors once defined it, is simply "a stretch of DNA that does something." For our purposes, we'll think of that "something" as coding instructions for building a protein. Since the strand of DNA making up a chromosome is continuous, with many genes in a row, an individual gene is referred to as a *locus* (plural: loci) from the Latin word for "place," referring to its place on the strand of DNA. We have two sets of all of our genetic loci, one set inherited from each parent, and these two sets may not be identical. The differing versions are called *alleles*, from the Greek word for "other."

The MHC is a large family of genes found in vertebrate animals. Animals can have many MHC loci. In humans, MHC is historically referred to as human leukocyte antigen, or HLA, although it is biologically the same thing as MHC. There are at least 253 HLA loci in the human genome, all clustered together on chromosome 6. About 150 of these loci are protein-coding genes. Researchers have identified a grand total of over 3,200 possible HLA alleles in humans. Thus, a huge amount of variation is possible among individuals and between populations. Statistically speaking, you would be pretty unlikely to encounter someone who had the same genotype as your own unless they were a close relative. However, some alleles are more common than others, and there is a wide spectrum of similarity and dissimilarity.

The proteins coded by MHC genes help the body identify pathogens and other foreign invaders, targeting them for destruction. Researchers have divided these genes into classes based on the invaders that they target:

- Class I MHC molecules activate the body's immune response against intracellular pathogens, such as viruses. When the animal's own cell has been infected with such a pathogen, MHC is part of the process that triggers the cell to undergo apoptosis, a form of programmed cell death. Of the 253 MHC loci found in humans, 128 of them are class I genes.

- Class II MHC molecules target extracellular pathogens, such as bacteria, recruiting T cells to mount an immune response. Humans have twenty-seven class II loci. This group of genes seems to have the most interesting relationships with animal odor and behavior.

- Class III MHC molecules are poorly understood, but they seem to be involved in signaling between cells.

MHC class II genes code for a protein that is expressed on the outside of so-called professional antigen-presenting cells. This category of immune cells includes macrophages,* B lymphocytes,† and dendritic cells.‡ The surfaces of these cell types are covered with MHC proteins that are able to bind to peptides—fragments of proteins—from the body's own DNA (when uninfected) or from pathogens (when infected). A peptide that forms a bond with another peptide, often producing a signal to other cells as a result, is called a ligand (from the Latin ligare, "to bind"). The portion of the MHC gene known as the "peptide binding region" codes for MHC ligands and is the part of the gene where we see so much variation. When an infection occurs, the MHC ligands on these immune cells will bind to a small peptide from the infection, known as an antigen, and hold that peptide on the cell surface. In this way, the antigen is presented to other cells in the immune system, stimulating an immune response. The ligand can't bind to just any antigen. Proteins have intricate three-dimensional structures, so the amino acid sequence and subsequent folding patterns of the ligand determine which shapes the ligand can bind. We sometimes refer to this binding as "recognizing" a pathogen. If you have a diverse MHC genotype, that means you carry several variants of these ligands, enabling your cells to bind to a variety of antigens from different pathogens.

I am no expert in immunology. The immune system is incredibly complicated, and I want to describe it here as simply as possible to avoid the quicksand of jargon. So, I was frustrated when I found

---

* Macrophages (from the Greek for "large eaters") are white blood cells that engulf and digest particles and cells that are deemed outsiders.

† B lymphocytes (sometimes just called B cells) are white blood cells that secrete antibodies to fight pathogenic bacteria and viruses.

‡ Dendritic cells are immune cells whose primary function is to communicate between parts of the immune system.

that most definitions of antigens seem obtuse. The clearest explanation I discovered came not from a text but from my husband, who stated succinctly, "So, an antigen is an irritant?" Well, yes. An antigen is something that irritates your immune system, stimulating it to respond by producing antibodies and inflammation. This description reminds me of how oysters secrete nacre to surround an irritating particle, ultimately producing a pearl. While writing this section, I learned that it's not actually grains of sand, as most people believe, but parasites that stimulate the oyster's response, making it an even better analogy.

MHC was first discovered by researchers who wanted to understand why someone's body might reject a donated organ. That's where the name comes from—the "histo" in *histocompatibility* refers to body tissues. They discovered that if the donated organ is from a person with an incompatible genotype, the antigens presented by MHC ligands on the recipient's cells activate an immune response to attack the foreign organ. In contrast, if the donor has a similar MHC genotype, the recipient's body acts as if the organ is its own. Thus, MHC molecules are often described as being part of the body's "self-recognition" system.

THE STINKY T-SHIRT STUDY

In the 1970s, researchers who were breeding strains of laboratory mice for immune studies happened to notice that the mice were more interested in mating with mice that had different MHC genes than their own. This serendipitous observation prompted the first controlled study to test whether the mating decisions of mice were determined by MHC genotype. The study found that, when given the choice, male mice did indeed mate more frequently with dissimilar females. Further studies on MHC-based mate choice in lab mice throughout the next few decades continued to support this conclusion. In particular, it became clear that these rodents were attracted to the *scent* of potential mates with different MHC genes. Researcher Dr. Claus Wedekind, at the University of Bern in Switzerland, began to wonder if this phenomenon was restricted to lab mice, or if it was more widespread. Was it possible that this law of attraction also applied to humans?

In the early 1990s, Wedekind and his collaborators recruited college students to test whether MHC genotypes affected how attractive men's body odors were to women. To prepare the scent samples, forty-four men were asked to wear the same white T-shirt to bed two nights in a row. The researchers provided unscented soaps and detergents and asked them to avoid wearing deodorant; they also asked the men to avoid smoking, eating spicy food, and engaging in other potentially smelly activities.* Soon after the T-shirts were collected from the men, forty-nine women were asked to smell the shirts and rate the odor for pleasantness and sexiness. To prepare their scent perception, the women were asked to use a nasal spray regularly during the two weeks preceding the experiment to ensure the health of their nasal mucous linings. Each woman was also given a copy of Patrick Suskind's 1985 novel *Perfume*, presumably with the idea that reading a book that focused heavily on scents would increase the attention they paid to odors.

On the day of the experiment, the women were each presented with six T-shirts in individual plastic bags. Three of the shirts were from men with MHC genotypes similar to their own, and the other three were from men with dissimilar MHC genotypes.

Echoing the results from the lab mice study, the women reported that they preferred the scent of T-shirts worn by men with dissimilar MHC genotypes. Perhaps humans, like other animals, were unknowingly engaging in disassortative mating—choosing a mate different from themselves—specifically at MHC loci. The biological advantage of this mating pattern is that the resulting

---

* Other activities that the men were asked to avoid included sexual activity and alcohol consumption. Wedekind noted in a magazine interview that it was challenging to recruit enough men for the study.

offspring have diverse MHC genotypes, better preparing them to fight off and survive infections and diseases, and generally increasing their chances of survival. The media found this result enticing, leading to much speculation about human "pheromones" and how they affect our relationships.

When I first learned about the idea that mate choice could be determined by MHC genotype, I, too, was fascinated. MHC and most of its effects are completely invisible to us, so could MHC-based mate choice explain decisions that otherwise seem to make no sense? Are our preferences determined more by biology than we realize?

Before we get too excited, there are several caveats to consider. First of all, not all women showed this preference. In fact, one group of women in the study—those who were taking hormonal contraceptives—actually preferred the scent of men who were *similar* to themselves. Hormonal birth control generally works by maintaining steady levels of estrogen and progesterone, simulating conditions during pregnancy and thereby preventing the hormonal fluctuations that trigger ovulation. Perhaps a woman's scent preferences are affected by her current hormonal state. Some researchers have speculated that this shift in preference might have its own evolutionary advantage: pregnant women would benefit from associating with family members who have similar MHC to their own, rather than expending energy seeking out potential mates with whom, at that time, they could not reproduce.

It's also worthwhile to question whether an attraction to the scent of an unwashed T-shirt really tells us anything about who we might realistically choose as a mate. To get a better understanding of possible MHC-based mate choice in humans, some researchers have studied the genotypes of actual couples. The genetic database HapMap II* includes genotypes from four geographically

* "HapMap" is short for Haplotype Map. A haplotype is a set of DNA variations that tend to be inherited together, so they may include multiple linked genotypes.

diverse populations in Africa, Asia, and North America. In two of the populations in this database, the Yoruba in Nigeria and a Mormon community in Utah, samples were collected from married couples, thirty in each population. Researchers found that mating patterns with respect to MHC were not the same in these two groups. The white Utah couples did show disassortative mating at MHC loci, and more so at those loci than at other points in the genome. However, in the Yoruba couples, there was no pattern of similarity or dissimilarity at MHC loci at all.

Pheromone researcher Tristram Wyatt also questions whether preferences for armpit odors tell us anything about human mate choice. Armpits are generally not where mammals sniff when assessing the scent of a potential mate. Instead, they are much more interested in the scents around the genital region, an area that is rich with scent glands. Of course, crotch-sniffing is not considered polite in humans, and, since we are bipedal, the armpits are much closer to our noses. Perhaps armpit odor is the first thing we notice about another person. However, not all human populations produce the same kind of armpit odor. In particular, smelly armpits are uncommon in the Asia-Pacific region. In these populations, a mutation in a gene called ABCC11 stops the production of compounds that feed the bacteria which emanate that distinctive underarm odor. Interestingly, the authors of one study of this gene hypothesized that a preference for low-odorant mates was the driving force behind the spread of this mutation throughout these populations.

It's tempting to jump to conclusions based on these intriguing studies. After all, the idea of MHC-based mate choice intuitively seems to explain so much. Some eager entrepreneurs have even started companies offering MHC-based dating services, although a number of them seem to have gone out of business already. But data from humans and other animals suggest that this phenomenon may involve many facets that we have yet to discover.

MHC IN BIRDS

Although the major histocompatibility complex is found in all vertebrate genomes, for a long time we didn't know much about MHC in birds. With earlier DNA sequencing methods, it was tricky to isolate individual MHC loci. The first studies of avian MHC were in domestic chickens, and for a while it was assumed that chickens were representative of all birds.

Domesticated chickens appear to have only nineteen MHC loci, only two of which are class II MHC loci. This repertoire is much smaller than what we've seen in mammals (recall that humans have at least 150 protein-coding MHC loci). Researchers hypothesized that birds had a "minimal essential MHC," with only just enough required to keep the immune system functioning properly.

This hypothesis was excellent news for my younger self, who, as a naive new postdoc, thought that sequencing MHC genes for juncos for the first time would be a piece of cake. Other early studies of avian MHC had also found evidence of low numbers of loci, usually three or four (three in house sparrows, four in great reed warblers, three in scrub jays, and three in red winged blackbirds). How hard could it be?

I soon found out. In fact, I toiled away over confusing, garbled sequences for a couple of years. I didn't understand what I was doing wrong. Over time, I grew increasingly exasperated. In fact, it was my frustration over this very situation that occasioned the fateful conversation in the cafeteria line that altered the trajectory of my career, kicking off my journey into the world of avian odors.

In retrospect, my problem was that I was simultaneously sequencing multiple alleles from duplicated loci that had different sequences. *Gene duplication* is a mutational process by which a stretch of DNA is accidentally included twice during *meiosis*, a type of cell division that creates eggs and sperm, each of which have half of our genetic information, neatly packaged to give to the next generation. If the duplicated gene is complete and fully functional, it usually causes no problems. With two copies of the functioning gene, one copy can accumulate new mutations over generations, so that the genome can retain the function of one while adding functions to the new copy. This extensive duplication in the avian genome is what made MHC difficult to study in passerines until the advent of next-generation sequencing techniques. The traditional DNA sequencing method I was using at the time, called Sanger sequencing, targets a single gene or stretch of DNA at a time. Isolating that gene is difficult to do when there are multiple, very similar, copies. Instead, next-gen methods sequence a large number of DNA fragments in parallel, allowing the entire genome to be sequenced all at once and successfully capturing and identifying duplications. Indeed, with the new and improved sequencing technologies, evidence soon emerged indicating more extensive gene duplication than had been suspected in passerine birds, including the zebra finch and the collared flycatcher. Using these newer methods, researchers identified at least twenty and possibly as many as thirty-one MHC class II loci in the common yellowthroat (Geothlypis trichas), compared to twenty-seven in humans.

It is not clear why chickens, and some closely related species, have such a different pattern from passerine birds, but it could have something to do with their chromosomal structure. In chickens, all of the MHC genes, plus some other immune-related genes, are clustered together on chromosome 16. Like many other birds, chickens have more chromosomes than mammals (thirty-nine pairs, compared to twenty-three pairs in humans). Many of these chromosomes are tiny microchromosomes, which contain much less genetic material than the larger chromosomes. Chromosome 16 is one such microchromosome. Many birds carry three, rather than just two, copies of chromosome 16 and other microchromosomes, a condition known as trisomy.

In humans, most trisomies result in miscarriages. There are only a few that are not fatal to the developing embryo, including sex chromosomes (XXY, XYY, and XXX are all known to occur) and trisomy 21, more commonly known as Down syndrome. In chickens, microchromosome trisomies do not appear to result in complications, perhaps because they contain less material and thus affect fewer traits when an animal carries an extra copy. In this way, chickens can have more variation in the number of MHC alleles that they carry, and they can actually carry more copies of them, while still maintaining a very small number of loci.

WHAT DOES MHC SMELL LIKE?

Having established the basics of MHC—what it is and how it works—the question remains, "What's smell got to do with it?"

Some evidence suggests that the MHC ligands themselves— those peptides on the outside surface of the immune cells that bind to antigens—can be detected by chemical senses. MHC ligand molecules have been found in urine and sweat, and animals, including mice and stickleback fish, appear to be able to detect these compounds. Unlike the airborne, volatile compounds we typically think of as making up scents, these heavy protein-based molecules are delivered in liquid solution. Thus, some researchers believe MHC ligands are detected not by receptors in the main olfactory system but instead in the accessory olfactory system, which includes the *vomeronasal organ* and its receptors in those species that have this "second nose."

The vomeronasal organ* is located in the nasal cavity, just above the roof of the mouth. This organ plays an important role in detecting protein-based pheromones in most tetrapods ("four feet"), the subset of vertebrates that includes amphibians, reptiles, and

---

* So named because it sits near the vomer bone in the nasal cavity. The bone is named after the Latin word for "plowshare," which the bone's shape resembles.

mammals.* If you have observed a dog intently sniffing another dog's butt, or the urine another dog has left behind on the ground, you may have noticed that the dog is physically touching those surfaces with its nose. That's because some of the information the dog is getting is not from volatile odors but from heavier, liquid substances that it is investigating with its vomeronasal organ. The vomeronasal receptors in the organ function much like olfactory receptors, binding to specific compounds and sending the information to the accessory olfactory bulb, which sends further information to other parts of the brain, activating specific responses to certain signals.

This hypothesis may explain one reason birds were initially ignored in studies of MHC-based mate choice. Although they are part of the tetrapod group of animals, birds do not have a vomeronasal organ, which means they would be unable to detect such heavy molecules using chemoreception.

There is another group of animals that appear to lack a functioning vomeronasal organ, but whose ability to sense MHC is highly interesting to us: humans. Although most mammals do have this system, it seems to be absent in Old World monkeys, apes, and humans. Human fetuses have a structure that resembles vomeronasal organs in other animals, but many studies suggest that this structure either does not exist or is nonoperative in adult humans. The question of whether vomeronasal function occurs in humans is debated by some scientists, though many of the claims that it does function are controversial.

A functioning vomeronasal organ may not be required for sensing MHC, however. There are several hypotheses about how humans might detect these genes using their sense of smell. The MHC molecules themselves may break down into smaller volatile

---

* Birds are also tetrapods, since they have two forelimbs (wings) and two hindlimbs (legs).

compounds present in urine and sweat. Peptides in urine could be altered by MHC molecules, and the metabolites produced as a result could be detected. Perhaps a fragment of MHC molecules actually carries volatile compounds.

But another intriguing pathway for MHC genotype to affect an individual's odor is through the regulation of an animal's microbiome. Because MHC class II ligands target antigens from extracellular invaders, including bacteria, they exclude particular species from the microbiome by activating the immune system against them. An individual's MHC genotype at class II loci affects which bacteria are excluded—and which may be tolerated. If MHC genotype affects the presence or absence of particular odor-producing bacteria, then the resulting odor should be a reflection of the host's underlying genotype.

We know from studies of the gut microbiome that MHC regulates the bacteria found in animal digestive tracts. In three-spined sticklebacks, researchers found relationships between specific MHC alleles (versions of genes) and the abundance of certain bacterial families in the gut. For example, if a stickleback had the MHC allele P8, it was far more likely to have bacteria from the family Legionellaceae. Even more interesting, the fish with more divergent MHC genotypes had less diverse gut bacterial communities. These fish, which display more diversity within their own MHC genotype, have a greater variety of MHC ligands, and consequently must exclude more bacteria from their microbiome than fish with lower MHC diversity. To put it simply, the more diverse your MHC, the more bacteria your immune system can identify and kill, resulting in a less diverse microbiome.

In humans, MHC genotype influences how susceptible a person is to developing celiac disease, an autoimmune disorder in which eating gluten can lead to damage in the small intestine. When comparing the gut bacteria of newborn babies, a study found that those with MHC genotypes more closely associated with developing celiac disease later in life had higher proportions of certain groups of bacteria in their guts, in particular *Bacteroides* and *Prevotella*. In laboratory mice, researchers found that MHC genotype was directly related to immunoglobulin A (IgA) responses to commensal bacteria in the gut. IgA is the source of the antibodies involved in attacking extracellular microbes.

If MHC has such an important effect on gut bacteria, we might expect it to have a similar effect on other parts of the body as well—such as the skin. In humans, research into the underlying causes of eczema (also known as atopic dermatitis) and psoriasis, two common chronic inflammatory skin diseases, suggests that MHC-based immune responses to antigens from the skin microbial community may be involved. In amphibians like the hellbender salamander, an individual's MHC diversity affects the diversity of their skin microbial community, which can be especially important in amphibians. Their skin is typically covered in a mucus—an excellent environment for microbial growth—and many populations are susceptible to fungal infections like the chytrid fungus, which can be fatal.

We are only now beginning to extend these ideas to uropygial microbiome communities. French researcher Sarah Leclaire found that blue petrels who were more similar at MHC loci also had more similar feather microbiomes. Furthermore, blue petrels with high MHC diversity had less diverse feather microbial communities, echoing the finding that sticklebacks with higher MHC diversity had less diverse gut microbiomes. At the time of writing, I am working with Joel Slade and Kevin Theis to analyze data from a recent junco experiment to test whether there are relationships among MHC genotype, uropygial bacterial communities, and preen oil volatile odor. Perhaps by the time you are reading this chapter, we'll know more.

So, what *does* MHC smell like? No one really knows, but I suspect it might smell different to everyone.

MHC AND SCENT PERCEPTION

When thinking about odors, it's important to remember that everyone's sense of smell varies slightly. Humans have about four hundred different olfactory receptors, which may allow us to recognize over one trillion different odors. Different combinations of alleles at our olfactory receptor loci cause us each to perceive odors uniquely.

Olfactory receptor genes are found throughout the human genome, on most of our chromosomes. However, they are not evenly spread throughout the genome. They tend to occur in clusters, which means that the genes within each cluster are more likely to be evolutionarily linked. Perhaps most exciting in the context of this chapter, one of the largest clusters of olfactory receptor genes is found on human chromosome 6—the same chromosome where our MHC genes are located. Similar linkages have been discovered in the mouse genome and the chicken genome. Because MHC genes and olfactory receptor genes tend to be linked, MHC might even influence how we sense each other's odors.

What does it mean for genes to be linked? When we pass our genes on to our offspring in the form of chromosomes, the copies they receive do not stay exactly the same between generations. When meiosis occurs in a future parent, the chromosomes don't line up according to which grandparent contributed them, nor do they necessarily represent each grandparent equally. In effect, the chromosomes we pass on are not exactly the same ones we inherited, either.

When the chromosomes line up just prior to the cell splitting and pulling them away from their matching copy, they can exchange genetic material in a process called crossing-over. Because the matching chromosomes have the same genes, even if they are different alleles, the DNA sequences are mostly identical. This

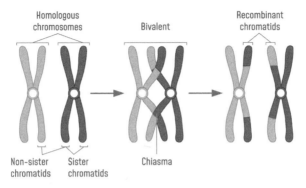

Crossing-over between chromosomes during meiosis.
Fancy Tapis / Shutterstock

similarity makes it easy for them to stick to each other and trade matching (or mostly matching) chunks. The farther apart two genes are on the same chromosome, the more likely they are to be split apart during crossing-over events. But genes that are physically located close to each other on the chromosome tend to stay together, and they are likely to be inherited together over many generations. Such genes are considered linked.

Thus, if particular olfactory receptor genes are linked to particular MHC genes, it could mean that individuals with specific MHC genotypes would also have associated olfactory receptors that result in sensitivity or preference for particular odors. This linkage could be one way that preferences for individuals with different MHC genotypes than yourself, for example, could come about.

We do not know whether one's MHC-associated olfactory receptor types really do affect the way you perceive odors. However, at least one study has found that human preferences for nonbiological odors, such as perfume, do correlate with a person's own MHC genotype. Manfred Milinski, an evolutionary biologist known for his work on MHC and mate preferences in fish (particularly the previously referenced sticklebacks), and Claus Wedekind (perhaps best known for the "stinky T-shirt study" mentioned earlier), teamed up while they were both at the University of Bern for a quirky study of perfume preferences. They recruited 137 male and female college students and laboratory assistants, who agreed to be genotyped at their MHC loci and to smell and rate thirty-six different perfumes. They were asked to smell the perfumes and rate their pleasantness in two contexts: first, as a scent they would or would not like to wear themselves, and second, as a scent they would or would not like to smell on a potential partner. The researchers found strong correlations between MHC genotype and which scents people would like to wear themselves. However, there was no relationship between MHC and the scents they would like to smell on potential mates.

It's not clear how we should interpret the results of this perfume study, except to note that there does seem to be a relationship between our MHC genotype and how pleasant we find certain odors. We don't yet know whether the olfactory receptors linked to MHC loci are influencing such a preference. But it is certainly an interesting hypothesis that could explain how we might go about detecting whether a potential mate is similar or dissimilar to us at MHC—and whether we find them attractive.

DO OPPOSITES ALWAYS ATTRACT?

So far, we have focused on the phenomenon of animals preferring to mate with others that are dissimilar at MHC genes. We know that the primary advantage of such disassortative mating is that it ensures offspring will have a diverse MHC genotype, strengthening their ability to detect and fight off diseases. It's easy to take this a step farther and assume that the best mate must always be someone who is the most different from you at MHC. But is that true?

In a study of wild house sparrows, researchers found that males that were too different from the females in the population—those males that didn't match the females at any MHC loci—failed to find a mate at all. Stickleback fish also avoid mating with those that are too different, instead opting for a mate that has a medium amount of difference from them, resulting in offspring with "optimal," rather than maximal, diversity. Sticklebacks may also achieve this optimal compromise by choosing a mate with a different level of diversity than their own: females who have low MHC diversity in their own genome prefer to mate with males who have high diversity, but females with high diversity instead prefer males with low diversity.

There may be a very good reason that we don't see all animals always choosing the most MHC-dissimilar mates: too much MHC diversity may actually be bad for their offspring's health. One hypothesis suggests that when an organism's immune system has an overabundance of different MHC molecules, it may begin attacking its own cells. Studies in mice and humans have found associations between MHC genotype and autoimmune diseases (for example, rheumatoid arthritis, multiple sclerosis, and type I diabetes). Several of these studies have indicated that interactions between multiple MHC loci can increase an individual's susceptibility to these diseases.

Another frequently observed MHC-related mate preference is for diversity *within* a potential mate's genotype, regardless of the level of similarity between the potential mates. In addition to avoiding males who were too different from themselves, female house sparrows avoided males who had low MHC diversity. Perhaps the scent of someone with a healthy immune system is very attractive to everyone. In this scenario, rather than comparing a potential mate's genotype to your own via their scent, you would simply be attracted to someone who already has a diverse genotype. After all, a mate preference for a diverse genotype would also increase the likelihood that your offspring would have desirably diverse genotypes.

Some birds do not show any evidence for MHC-based disassortative mating, nor do they exhibit any preference for diversity at these genes. In a study of Magellanic penguins (*Spheniscus magellanicus*), the similarity between mated males and females was no different than expected by chance. However, this population of penguins had very high levels of diversity, and it was very rare for any two penguins to share genes. In such a diverse population, any mate this monogamous bird chooses is likely to have a sufficiently different genotype from its own, and statistically we would not be able to detect whether the animals were choosing mates on that basis. Nor was there evidence for disassortative MHC-based mate preferences in the great snipe (*Gallinago media*)—though in this long-beaked, stocky bird, males with particular MHC genes seemed to be more successful than others, suggesting that females may prefer males with specific MHC genes rather than diversity. It's possible that those specific variants are more resistant to common diseases or parasites, so that the males carrying them are healthier and thus more attractive.

The context in which mate choice occurs—in particular, the genetic makeup of the population—matters in terms of whether MHC is important at all. Recall that no MHC-based mating patterns were observed in Yoruba couples in Nigeria, while white Mormon couples in Utah showed disassortative mating. Why might these populations have such different preferences? An important difference between these two groups is the level of genetic variation present in the entire population. Groups that have descended from a small number of immigrants and that have created insular subcultures tend to have low levels of genetic variation and a higher risk of inbreeding.

The Hutterites are a group of Anabaptists, originally from the Tyrolean Alps, that settled in North America in the 1870s. Members of this ethnoreligious group live in rural colonies in the US upper Midwest and in Western Canada. In a study of MHC in

Hutterite colonies in South Dakota, the approximately one thousand people sampled were descended from only sixty-four ancestors, resulting in high relatedness in the community. Even among spouses, the average relatedness is at a higher level than that of first cousins once removed. Because of this genetic bottleneck, the population has low MHC diversity overall. However, given the limited level of diversity in the local population, married couples had a greater number of MHC genotype mismatches than would be expected, suggesting that there was a greater degree of disassortative mating occurring in this population than would have arisen by chance alone.

Dr. Manfred Milinski points out that a preference for MHC-dissimilar mates may only be a "best-of-a-bad-job" rule. In other words, if you can't find a mate with the best MHC genes (in this case, Milinski is considering genotypes that provide resistance for specific parasites), then the next best thing you could do is pick a dissimilar mate. Thus, in a population with low MHC diversity and a high risk of inbreeding, as in some human populations like the Hutterites or in animal populations that are isolated on islands, it would be adaptive to have a preference that maximized the MHC distance between you and your mate. On the other hand, if your population is currently being ravaged by a specific disease or parasite, then your offspring would be better off if you chose a mate with MHC alleles that made your offspring resistant to those diseases.

Research on MHC-based mate choice in humans continues. In fact, the very same week I finished the first draft of this chapter, a new study was published, the largest one on this topic I've seen so far. Researchers genotyped 3,691 German couples at six MHC loci. They found no difference in dissimilarity between the married couples compared to the dissimilarity between two randomly chosen people in the sample. Furthermore, the couples in which the women had been taking hormonal birth control when they began dating (64%) were not more or less similar at MHC than the other couples—despite the "stinky T-shirt" study's finding that women taking hormonal birth control did not show a preference for the scent of MHC-dissimilar men, while the other women did. It is, of course, a distinct possibility that odor preference in a laboratory study may not be a good proxy for understanding mate choice.

Many questions remain about whether MHC influences human mate choice. For example, if MHC does not influence marriage, what about choice of partners for extramarital affairs? And what about same-sex partners? Do external circumstances, such as the presence of pathogens, influence the attractiveness of certain MHC alleles? Research thus far suggests that MHC-based preferences can be highly context dependent, and we've only begun to consider a few of those circumstances. Despite all of the difficulties in studying mate choice and MHC in humans, I suspect researchers will continue to find the topic irresistible for many years to come.

FINALLY, SOME RESULTS

I did, eventually, succeed in sequencing MHC loci in dark-eyed juncos. I focused on the juncos from the two populations in San Diego County, California, that I had studied with Jonathan Atwell: the quickly evolving population that had taken over the University of California San Diego campus and their possible ancestral population in the Laguna Mountains. These groups present us with a "natural experiment," in which different populations are organically exposed to different conditions, giving us an opportunity to study evolution in action. Genetic evidence suggested the UCSD population had been founded by only ten or fewer breeding pairs, resulting in a loss of genetic variation. I wondered whether this bottleneck had also resulted in a loss of variation at MHC loci—and whether that would make them more likely to choose MHC-dissimilar mates.

It took a few years of troubleshooting, but I was able to over-come the technical problems presented by gene duplication and obtain MHC data using an ABI 3730 DNA Analyzer—a nice machine at the time, but it still used traditional Sanger sequencing rather than next-generation methods. I'll spare you the nitty-gritty details, but DNA sequencing involves two major steps, both of which involve pipetting tiny amounts of liquid into tiny tubes. The first step is called PCR, short for *polymerase chain reaction*, and it amplifies—makes copies of—the gene of interest by artificially forcing the DNA to duplicate itself, but only between the two ends of the piece of DNA you want to amplify. You can specify these ends using a pair of primers, which are just short pieces of DNA matching these ends in your sample's DNA. Once you have amplified your gene, you should have a tiny tube full of clear liquid that contains copies of the gene you are sequencing. In the second step, a sequencing reaction tags each nucleotide in the DNA with a fluorescent molecule. Finally, you take the results of the sequencing reactions, load them into a small rectangular plastic plate with ninety-six tiny tubes, or "wells," and pop the plate in the sequencing machine, which will run the liquid through tiny capillaries and detect the order of the fluorescently tagged DNA molecules using lasers. The whole process involves a lot of faith in those clear liquids and invisible reactions.*

I sequenced four MHC class II loci, which produce the MHC molecules responsible for detecting extracellular invaders like bacteria. I found that the UCSD junco population and the Laguna Mountain junco population had essentially the same levels of variation at MHC loci. How had the UCSD population maintained high variation at MHC despite losing diversity at other places in the genome? This unexpected finding may be due to selection pressures

---

* Many of the labs where I've worked have had little altars next to the PCR machine where grad students made offerings to the PCR gods.

of disease. Atwell found that UCSD birds had more feather mites than Laguna Mountain juncos, suggesting that the urban birds were encountering more parasites in their new environment. Selection pressure from parasites could keep variation at MHC class II loci high in this population, so that only those with high MHC diversity remain healthy and leave more offspring. However, later work by Rachel Hanauer (another graduate student in the Ketterson lab at Indiana University) compared eleven different junco populations in California and found no difference in parasite prevalence between urban and nonurban populations.

Because of all the technical difficulties I encountered, the conclusions that I published in 2012 were a far cry from the research I had set out to conduct in 2006. By the time I finally published that paper, I was so sick of MHC that I swore I'd never study it again. Obviously, that was a promise to myself that I failed to keep, mostly thanks to Joel Slade joining me at MSU and reviving my interest in these genes.

I had planned to return to southern California with Joel, where together we would sample these populations again to learn more about the relationships among MHC, odor, bacteria, and mate choice. I wanted to know whether the odor differences I had previously observed between these two populations were due to distinct microbial communities in their uropygial glands. I suspected that the UCSD birds, living in close contact with humans and their garbage, likely had picked up some microbes that weren't common in the juncos that lived in less-disturbed forested areas. Joel also hypothesized that the two populations could be subject to different diseases, beyond the parasites previously investigated, and that they would show changes in MHC genotype as a result. We had also planned to compare the MHC genotypes of mated pairs using new and improved methods that Joel had developed for the juncos.

Despite my MHC-fatigue, I still wanted to know whether mate choice could explain how the small UCSD population had maintained the same level of MHC diversity as the Laguna Mountain population. I speculated that, perhaps due to reduced overall genetic variation in the population, the UCSD juncos might rely on MHC for mate choice (much like the Hutterite and Mormon couples in human studies), while the more diverse Laguna Mountain birds did not. If UCSD juncos were mating disassortatively at MHC, then diversity would be maintained in the population. To test this hypothesis using the data I already had, I compared the MHC genotypes of mated pairs, calculating the genetic distance between the male and the female.

Alas, there was no difference in MHC dissimilarity of mates at UCSD and Laguna Mountain. However, these two populations did display an important dissimilarity in their mating behavior. UCSD birds engaged in extra-pair mating much less frequently than in other junco populations (although it wasn't absent entirely). I combed through my earlier paternity testing data from this population, and here, I uncovered a divergence. At Laguna Mountain, there was no difference in MHC distance in pairs with or without extra-pair young. But at UCSD, the pairs with significantly more similar MHC genes also had more extra-pair young. The reverse pattern was true for the pairs that had remained faithful to each other—they had more dissimilar MHC genes. Compared to pairs that were already MHC-dissimilar, perhaps those juncos with mates too genetically similar to themselves sought out extra-pair matings to ensure their offspring would have more diverse MHC.

Researchers have hypothesized that females mated to males of lower genetic quality might be more motivated to seek out extra-pair copulations. In the Seychelles warbler, females mated to males with low MHC diversity were more likely to seek out extra-pair mates. Female Savannah sparrows with MHC-similar mates also had more extra-pair young. Unfortunately, my preliminary analysis on the relationship between MHC similarity and extra-pair fertilizations in the UCSD and Laguna Mountain juncos is based on some shaky data and would not hold up to peer review at this point.

Joel and I still hope to do a more robust study on this topic. However, this work was put on hold due to pandemic restrictions. We are optimistic that we'll be able to embark on the next stages of this research in the near future. Other urban populations of juncos have started appearing, especially on the West Coast, but I still have a sentimental attachment to the San Diego population—the original urban juncos.

CHAPTER 8

. . . .

GIRL POWER

*Choosing the right scent was as important as choosing the
right dress—you wanted the boy to like both. This is the perfume I wore when your father and I were courting. We used
to meet at a rose garden on the hill south of town, and I had a
terrible time finding a fragrance that wouldn't be overpow-
ered by the flowers. When the wind rustled my hair, I would
give him a look as if to ask whether he'd noticed my perfume.*
—YOKO OGAWA, The Memory Police

THE SECOND SEX

When we think about visual, auditory, or chemical signals in the
context of bird behavior, we think of males first. The ways they
appeal to the senses to acquire mates also win human admirers.
We think of their lovely songs, their come-hither bright feathers,
and the dances or battles males employ to demonstrate their de-
sirability and dominance.

In contrast, we expect all female birds to be smaller, clothed in
drab colors, and relatively quiet, a pattern seen in a wide variety
of animals. These characteristics help females to blend in with the
environment and escape attention from predators so that they can
safely attend to the critical role of raising offspring. While some

species fit this template, it doesn't follow that females have no need to attract mates or compete with other females.

In fact, the females of many species are just as ornamented or armed as the males, yet it is common to dismiss those examples as exceptions to the rule. Some researchers have suggested that showy colors or ornaments in females—such as in toucans and macaws—are simply inherited from their male ancestors as a sort of side effect when there is no evolutionary cost to having them.* However, in more recent years, some scientists have begun to think more critically about female ornamentation and started testing ideas about whether there may be selection on the basis of these traits in females, as well. Just as in males, striking visual cues can function as a signal of quality, which undoubtedly makes them useful in competition with other females for resources and for attracting the opposite sex.

Although the study of avian chemical communication is still quite new, research in the area already seems to be constrained by the same assumptions that long limited investigations of other communicative modalities like song and plumage ornaments. The assumption that preen oil odors should function primarily as attractive male chemical signals is so entrenched that it seems to have prevented some researchers from noticing that the females actually produce stronger odors, even when their own data clearly demonstrated this pattern. In a study of budgerigars (the parakeets that are commonly kept as pets), one group of researchers noted that, in two-way choice tests, females preferred the scent of males over that of females. The researchers then went searching for a male pheromone that apparently accounted for this preference. They focused on three compounds that make up a larger proportion of male odor compared to female odor: octadecanol, nonadecanol, and eicosanol. Together, these compounds made up, on

---

* The same argument has been used to explain the existence of the female orgasm.

average, about three-quarters of each male's total volatile profile, dominating his odor and creating what the researchers deemed a "male signal." Yet, the data presented in this same paper clearly show that females produced even greater amounts of not just these three compounds but at least seventeen additional compounds, as well. The only reason those three particular compounds seemed less important in female budgerigars was because females produced more odor overall, so those compounds made up a smaller proportion of their total odor profile.

I must admit that I, too, fell victim to this way of thinking. You may have noticed that in this very book I have written many words on how male birds might use odors to signal their quality to females or to other males. This bias existed despite the fact that, early on in my career studying avian smells, I noticed females tended to have larger uropygial glands, and in my datasets, females produced higher quantities of preen oil volatile compounds. Julie Hagelin, a pioneer in avian chemical communication, had noted the same thing, and a year or two after we first met, we excitedly decided to write an article together about why females might produce more preen oil than males. But at the time—way back in 2011—there were only a handful of birds for which the chemical compounds had been quantified, and it was impossible to know if our findings were indicative of a general pattern. Life soon intervened, and the project fell by the wayside.

However, as more studies on avian chemical communication were published, the pattern held, and our idea continued percolating in the background. In 2018, Julie and I were reunited when we were both invited to a special symposium organized by Barbara Caspers at Bielefeld University in Germany. The symposium was called "Social Olfactory Communication," and although the primary focus was on birds, the program also included mammal researchers sharing their insights. I was thrilled to finally meet many of the people whose research I've described in this book.

Julie's talk revived our idea about female chemical signals being potential drivers of differences between the sexes. The presentation went over well—the idea seemed to be something that others had not considered before, and the other chemical ecology researchers were interested in this new way of looking at the phenomenon. Over a hotel buffet breakfast on the second day, Julie and I decided to get back to work. We set out to compile all of the existing data we could find on sex differences in production of odors in birds, and to try to hypothesize what female odors, in particular, might be used for.

Confirming our initial impressions, we found that that, across bird species, whenever there was a difference between the sexes, it was females, not males, who topped the measurements of every aspect we examined. Females had larger uropygial glands and either higher concentration or higher diversity of preen oil volatile compounds, as well as more varied bacteria. In many cases, these sex differences only appeared when the birds were in breeding condition.

In both sexes, the uropygial gland tends to increase in size during the breeding season, since birds produce more preen oil when they are reproductively active. But here, too, the increase is relatively greater in females. I thought back to my 2011 field season studying the behavior of pink-sided juncos at Grand Teton National Park, when my field assistant and I measured the uropygial glands of all of the juncos we caught as part of our regular measurements and sampling. Uropygial glands are tiny, just a few millimeters in each dimension, so we had used calipers with a digital readout that could give minutely precise measurements. Measuring the height, length, and width of the glands, then multiplying these numbers to get a rough estimate of gland volume, we found that all the females had larger gland volume than the males—nearly 20% larger. For years, I wasn't sure what to do with that information, but once Julie and I began compiling our data across species, it became very relevant.

Other researchers noted the same pattern in other bird species. Female great tits have about 30% larger uropygial glands than males during the breeding season. Both female and male zebra finch uropygial gland size increases during breeding, peaking when the chicks hatch—and during this peak, females have significantly larger glands than males. And it wasn't surprising that female Eurasian hoopoes—those birds that produce preen oil full of antimicrobial-producing bacteria to protect their eggs—proved to have much larger glands than males when breeding. However, a couple of exceptions to this rule have been found, which may be related to important environmental adaptations. The common eider is one of the few species in which males have larger glands than females. This large diving duck breeds in the Arctic, making the waterproofing and insulating properties of its preen oil important for protection against the extremely cold water. Male eiders are larger than females and spend more time at sea and thus need to produce more preen oil to coat all of their feathers. Another exception was observed in a study on barn swallows, in which uropygial gland size was negatively correlated with feather-degrading bacteria. Because preen oil has antimicrobial and antiparasitic properties, birds with larger glands had fewer parasitic bacteria, regardless of sex. In this study, female barn swallows had more feather-degrading bacteria and smaller uropygial glands than males, though it's not clear why.

With these thought-provoking results on gland size in hand, Julie and I next compiled all of the data we could find from studies that quantified preen oil volatile compounds. We uncovered papers on a total of sixteen species of birds. Of those sixteen species, twelve displayed significant sex differences in odiferous compound concentrations. In eight of those twelve species, females produced stronger concentrations than males overall (including budgerigars, as noted above). We found only one species, the Bengalese finch, in which the bias emphasized males: there were two volatile compounds that were higher in males and only one that was higher in females. In the other three species that had significant sex differences, it was not possible to point to one sex or the other as having overall greater concentrations of compounds. Instead, males and females produced different profiles, where some compounds were higher in females and other compounds were higher in males. In addition to the studies describing twelve species in which females produced more powerful scents, we found studies on another four species of birds in which the specific volatile compounds produced or the diversity of microbes on their feathers or skin differed between the sexes. The males were not more diverse in any of those cases. Instead, females produced a greater number of unique compounds than males or exhibited greater microbial diversity.

Now we knew that females had larger uropygial glands and stronger and more diverse odors than males during the breeding season. The obvious next question was *why*? We had focused so much on how males might use chemical signals that it wasn't clear what aspects of female behavior might require them.

Around this same time, these questions about sex differences, and the bias in studying them, were also being raised in the field of mammalian chemical signaling. Holly Coombes, at the time a PhD student working with Jane Hurst and Paula Stockley at the University of Liverpool, noted that studies of mammalian chemical communication were strongly biased toward understanding how males use scent to attract mates and to compete with each other. Yet this androcentric focus neglected to address the multitude of ways that females use chemical communication. Coombes compiled an impressive review paper identifying the aspects of female social behavior where chemical signals were likely important and where we needed more research. The three areas she highlighted should sound familiar: mate attraction, competition with other females, and parental care. Julie and I agreed that there was no reason the same ideas that Coombes compiled should not apply to birds too.

THE FEMININE MYSTIQUE

The Parental Investment Hypothesis, posited by evolutionary biologist Robert Trivers in 1972, states that whichever sex invests more energy into its offspring will be choosier about who they mate with. In vertebrates, that is typically the females, who produce more energetically expensive gametes (eggs, rather than sperm), may incubate those eggs in a nest or gestate the embryo inside their own bodies, and often provide more parental care to developing young than males do. Meanwhile, members of the other sex will spend most of their energy competing with each other for access to mates, as we see in many male vertebrates. This idea of differential investment goes a long way toward explaining sex differences in behavior across many species of animals. Unfortunately, the concept also tends to be oversimplified, causing many well-meaning researchers to overlook the variation that occurs between species, and even within each sex.

In fact, reproduction is costly for both sexes. When males spend energy trying to attract females or fighting with other males, they have less energy available for finding food or watching for predators. They may suffer consequences, such as not obtaining enough calories to fuel any activities requiring sustained effort, poorer health and resistance to disease, and even death. Furthermore, in species where pair bonding and biparental care are the norm—as in many birds—males who seek extra-pair mating opportunities are leaving their mates alone while they search. The males' wandering, of course, leaves their female partners free to pursue their own outside sexual dalliances. In that case, what's good for the goose may not be good for the gander. As a result of his straying, he may ultimately lower his own reproductive success rate by expending his energy raising offspring that are not his own. To maximize benefits in the face of such costs, males should choose their mates wisely. A high-quality female will be more likely to raise surviving, healthy offspring, thus ensuring the male's genetic legacy. If males are also choosy, then females should be advertising their availability and quality to potential mates just as males do. So why not emanate a desirable perfume?

Scent can play an important role in synchronizing the timing of mating between males and females. Courting and attempting to mate with a female who is not in reproductive condition is a waste of a male bird's energy, while increasing his risk of being attacked by another male, or by a predator. And we know that constantly maintaining high levels of testosterone and sperm production is also costly. Chronically elevated testosterone can have detrimental effects on other aspects of the male's physiology, perhaps even directly decreasing life span. To maximize success, a male's testosterone levels need to fluctuate so that they are higher when it is time to mate and lower when it is not. It stands to reason that a signal from females indicating their hormonal and physiological state would help ensure that males expend their effort at the right time and with the right mate. The previously reported finding that male chickens were less likely to copulate with females whose uropygial glands had been removed strongly suggests that the scent of a reproductive female may have a priming effect on the male's body. A female's chemical signal could trigger a cascade of hormonal responses leading to an increase in testosterone, sperm production, and the male's desire to mate. This priming response has been observed in several mammals, including rodents, primates, and even humans, but it has not yet been studied in birds.

Seasonal patterns in scent production suggest that chemical signals in both sexes could be important in stimulating reproduction in birds. So far, all studies examining seasonal changes in production of preen oil volatiles have found that both sexes produce stronger odors during the breeding season. In females, there are also finer fluctuations that appear to correspond with ovulation patterns. In dark-eyed juncos, a female produces the strongest concentration of preen oil volatile compounds just before egg-laying, which might signal to males that she is ovulating and ready to mate. This pattern, like others, echoes scientific observations of many female mammals, who produce distinctive or more intense odors when they are in breeding condition to signal their receptivity. In several species of lemurs, for example, when females are in breeding condition, the composition of their genital and perianal secretions goes through distinct changes. Behavioral studies of many mammals, from domestic dogs to golden hamsters and giant pandas, have shown that males can distinguish between the scent of breeding and nonbreeding females, and that they are usually more attracted to the scent of breeding females.

Not only do female mammals produce particular odors when they are ready to mate, they also go to more effort to spread those odors around. Females scent-mark more frequently when they are sexually receptive. For example, although male giant pandas scent-mark trees at about the same rate year-round, about 80% of female giant pandas' scent-marks in one study were made during the breeding season. This behavior can be especially important for solitary species who need to find mates. In female tigers and golden hamsters, both solitary-living species, scent-marking peaks right before the females go into estrus, alerting the males in the area that there is a reproductive female nearby looking for a mate.

Of course, we humans are always deeply interested in determining whether the underlying biological drivers controlling behavior in other animals also apply to us. Whether we find these results illuminating or disquieting, there is no doubt that we will continue to explore the connections between scent and sex that we ourselves may unconsciously manifest. In a Finnish study using the same methods as the classic "stinky T-shirt" study, subjects were asked to smell T-shirts worn by women at different stages of the menstrual cycle. The subjects, which included both men and women, were asked to rate the scents for sexual attractiveness and intensity. Men found the scent of ovulating women most attractive. Interestingly, the women showed the same preference, although it was not statistically significant.

In addition to advertising their availability and sexual receptivity, females may use odor to signal their quality and compatibility to potential mates, just as males do. In ring-tailed lemurs, information about MHC genotype, including heterozygosity or similarity to another individual's genotype, is present in the odor of both sexes. Female odor can also reflect other aspects of quality, such as nutrition and vulnerability to disease. For example, male mice preferred the scent of females who were not infected by a nematode, and male meadow voles preferred the scent of females fed a high-protein diet over those fed a low-protein diet.

Few studies of birds have done a satisfactory job of testing male preferences for female scents. This lack is partly due to the difficulty of interpreting the meaning of such choices in a laboratory experiment, and partly because of the traditional research emphasis on female preferences for male scents. Even so, some researchers have documented scent-related preferences, especially with respect to MHC genotype. In a study of red junglefowl, males mated with females regardless of their MHC genotype—but they gave more sperm to MHC-dissimilar females. Although this study did not specifically examine differences in scent preference, we know that MHC genotype is thought to be detected via scent. More directly, Sarah Leclaire tested the preferences of both male and female blue petrels with respect to MHC similarity. She captured wild petrels and observed their reactions to preen oil odors from the opposite sex in a Y-maze. Unexpectedly, 75% of the females tested preferred the scent of MHC-similar males, but 100% of the males preferred MHC-dissimilar females.

We have much more work to do before we can confirm that female bird odor plays a significant role in stimulating males to mate, or in male preferences for females. But the studies that have been done so far provide provocative, if indirect, support for the idea. However, if we truly want to investigate avian olfaction in a more balanced way, it's crucial to remember that a female's behavior— and the possible role of her scent—involves much more than attracting males. Females also interact with each other.

GIRL FIGHT

Although aggression in females has been studied less than in males, we know that females must compete with other females for access to resources like food, territories, nesting sites, and mates, just as males do. Animals can reduce the risk of violent confrontation by advertising their competitive abilities through vocal, visual, or chemical signals. These signals allow potential competitors to assess their quality and decide whether it's worth the risk to engage or if it's better to walk—or fly—away. Several studies have shown that female birds flex through plumage ornaments, song, and even odor.

Although many bird species feature brightly colored males and dull-colored females (this phenomenon is known as sexual dichromatism, a type of *sexual dimorphism*), there are plenty of bird species where the females are just as brightly colored as the males. Some researchers have suggested that plumage monomorphism— when both sexes have the same feather coloration—occurs more often when both the females and males of a species are prone to aggressive displays. One such species is the capuchinbird, a large orange bird found in forests north of the Amazon river. Female capuchinbirds regularly fight with other females, and in particular they purposefully interrupt other females who are being courted by males. But even in sexually dimorphic species, females may be aggressive toward one another. As Kim Rosvall's research showed, female tree swallows can be very combative in their competitive search for nest cavities, swooping over or even pecking at intruders.

Although we usually think of males as the singing sex, females sing in nearly two-thirds of songbird species surveyed. Female northern cardinals sing as much as males, often in a territorial context, and they are just as belligerent toward territorial intruders as their male partners, posturing toward, flying over, or attacking their competitors. Female song sparrows primarily sing to defend their territories from female intruders, as do the female superb fairy-wrens of Australasia.

Rosvall and I were the first to examine the relationship between odor and aggression in female birds. We found that preen oil odor predicts hostile behavior in both male and female juncos. As described in Chapter 3, female juncos were presented with a caged female interloper on their territory. Females responded by swooping over the intruder's cage and generally spending time near it, which is threatening behavior. Just as in males, the female's overall odor profile was related to these antagonistic behaviors, suggesting that in both sexes a potential rival could use scent to gauge how confrontational a bird might become when reacting to an intrusion.

In addition to signaling aggressive intent toward potential intruders, there are other contexts where female odor might be useful in competition. One particularly interesting possibility occurs in cooperatively breeding species. In these species, the mated pair receives help raising their offspring from other adult, nonreproductive individuals in their group. Social groups in many cooperatively breeding species, such as the Florida scrub jay, include the breeding pair and their older offspring who have not yet left home to start their own families, instead staying in the group to help raise their younger siblings. Helping to raise your siblings at the expense of your own reproduction—either delaying it or eschewing it entirely—still increases your evolutionary reproductive success, since you share genetic material with your siblings and are pitching in to ensure those genes stay in the population. The theory describing this evolutionary tactic is known as kin selection, since natural selection is operating via relatives. However, helpers may also end up raising young that are unrelated to them. In the tiny, promiscuous superb fairy-wren, most offspring are sired by extra-group males,* and if the breeding female dies or leaves the group, her remaining older offspring assist their father's new mate in raising her chicks. It can still be advantageous to live and cooperate in a group like this if it increases your ability to defend resources, helping you to survive. Plus, there is the possibility that the nonbreeding adults could replace the breeding pair, either by fighting them or outliving them.

In many species of cooperatively breeding mammals, subordinate females do not even ovulate in the presence of the dominant, breeding female. One hypothesis suggests that the dominant females give off a pheromone that suppresses reproduction in the subordinate females, although more detailed work has not fully supported this idea. Common marmosets, a very small South American primate species, always give birth to twins, and the mothers need help from other group members to care for these relatively large babies. While studies demonstrated that subordinate females from the same group did not ovulate in the presence of a dominant female, further investigation showed that odors from a different, unfamiliar dominant female did not suppress ovulation in subordinates. So, instead of a "suppression pheromone" produced by all dominant females, subordinate females may

---

* During courtship—which apparently happens often, since up to 75% of all nestlings are sired by extra-pair males—superb fairy-wren males often present females that catch their attention with tiny yellow flower petals.

simply recognize the scent of the dominant female from their own group. Suppression of reproduction could come about through social behavior, and the scent of a dominant female could simply serve as a reminder of that behavior and its consequences.

In birds, the link between olfaction and reproductive suppression has never been investigated, but the idea is worth pursuing, especially given the fact that breeding female birds produce stronger odors than males. Some combination of inbreeding avoidance, competition, and reproductive suppression prevents subordinate birds from breeding in groups where the nonbreeding birds are related to the breeders. It may be simply that the scent of relatives inhibits reproduction as an inbreeding avoidance mechanism, as we saw in zebra finch females housed with brothers. However, inbreeding avoidance alone doesn't explain why subordinate females don't reproduce, even when an unrelated breeding male joins the group, as in the southern pied babblers that inhabit the savannahs of southern Africa. Subordinate birds of both sexes generally have lower levels of reproductive hormones than breeders in cooperatively breeding groups, and in some cases their reproductive physiology is less developed. A previously popular hypothesis to explain suppression in these birds was "psychological castration": breeding adults kept the subordinate adults stressed by behaving aggressively toward them. These high levels of stress would keep the subordinates' corticosterone levels high, and we know from other studies that high corticosterone can negatively affect reproductive hormones and behavior. However, when comparing adults in cooperatively breeding species, researchers have not found any difference in the corticosterone levels of the breeding pair and the subordinate adults, suggesting that some other mechanism is occurring.

Of course, there is ample evidence of the competitive power of scent in other species. In mammals, scent-marking by males is often deployed as a way to advertise one's competitive abilities toward same-sex competitors. Scent-marking by females has often been attributed to the need for communicating reproductive status to males, but perhaps there is more to the story. Although female ring-tailed lemurs scent-mark most frequently when they are in breeding condition, they actually scent-mark more frequently than males throughout the year. Males investigated both male and female marks, but females were primarily interested in the scent-marks of other females. Female ring-tailed lemurs are dominant over all of the males in their group, and among the females there is a strict dominance hierarchy. Females compete intensely for resources and may be quite antagonistic toward lower-ranking female members of their group. Lemur aggression ranges from warning signals like staring and huffing to all-out fights, complete with striking and biting.

Do birds have any behaviors that might be considered scent-marking? I have proposed that bill-wiping, when a bird rubs its beak against a branch or other object during an interaction with another bird, may be a sort of scent-marking. If residual preen oil is on the beak, then rubbing it against a surface could release odors that were trapped in the dried wax. When I studied this behavior in juncos, I tested whether it was more important in courtship or aggressive interactions —unfortunately, I only studied it in males. When I designed the study, I hadn't even thought to consider this behavior in females. Males used this display much more often in courtship interactions with females than they did in aggressive interactions with other males. However, in some of my behavioral trials, the resident female would also appear and behave aggressively toward the intruder in the cage, whether it was a male or a female. I did notice at the time that many of these females also bill-wiped; in retrospect I wish I had included them in the study. Understanding whether bill-wiping is truly some form of scent-marking (or, as I called it, an olfactory display) requires more research, and unlike in my study, future research really needs to examine female behavior, too.

MOTHER KNOWS BEST

In addition to playing an important role in attracting and synchronizing mating with males, and competing with other females, female birds may also produce more preen oil and stronger odors than males to support maternal care of offspring. We know that a junco mother's preen oil likely offers her nestlings protection from parasites when her feathers rub against them during incubation. We also know that Eurasian hoopoe females spread superpowered (and super stinky) antimicrobial preen oil on their eggs, where it is held by microscopic crypts (minute holes) in the shell's surface to guard against pathogens.

But is there a way in which just the scent of a mother's preen oil could protect her nestlings? Sure, if it could alter the smell of the nest, potentially masking it from detection by predators. In a number of shorebird species and ducks, preen wax composition changes when the birds are nesting. The waxes shift from lighter, more volatile monoester waxes to heavier, less odorous diester waxes. Because the diester waxes do not give off as many volatile compounds, it's thought that predators are less able to detect the breeding birds and their nests. As mentioned previously, in one study, a trained dog had a more difficult time finding diester waxes than monoester waxes, lending support to this hypothesis.

This idea is so compelling that when my labmates at Indiana University first measured volatile compounds in junco preen oil, they were surprised to discover that the birds produced much higher concentrations of volatile compounds in the breeding season. But chemist Helena Soini, a coauthor on that study and my collaborator in subsequent projects, noted that many of the volatiles that dominate female odor in the breeding season are also produced by plants. She suggested that perhaps these odors could help camouflage the scent of the nest. We've not yet tested this fascinating possibility, but it could open up a whole new avenue of research into olfaction's role in avian behavioral ecology.

Odor is also a key component of parent-offspring recognition. The scent of one's own offspring may induce a parent to provide care. In an extreme example, researchers painted a fruit-scented liquid onto ring dove chicks. The parents gave less care to these odd-smelling chicks, and the chicks were more likely to die. However, if the researchers cut the bilateral olfactory nerve of the parents, the effect went away, demonstrating clearly that the change in odor was the cause of the change in parental care. Perhaps this finding helped fuel the idea that the scent of humans would cause wild mother birds to abandon their nest—but, in reality, they never do, probably because the offspring's scent is stronger than any ephemeral odors left behind by interfering humans. Recognition of one's own offspring appears to begin very early, before the young hatch. For example, female zebra finches recognize the scent of their own eggs and prefer them to the eggs of other females.

Chicks and their parents have highly similar odors, as seen in juncos. At least some of a nestling's odor comes from its environment, including the brooding mother's preen oil and scents from the nest. Nests have individual odors that are influenced by the preen oil, saliva, and feces of the birds that live in them. When mother diving petrels and zebra finches recognize and prefer the scent of their own nest over other nests of their species, they may be responding to a scent that is at least partially their own; however, chicks do produce their own odors, beginning in the egg. Embryonic Japanese quail give off volatiles that convey information about whether the egg is fertile beginning as early as the first day of incubation, and the odors change as the embryo develops.

Interestingly, embryonic odors also signal sex differences in these quail. The same is true in barn swallows.*

Nestlings recognize the scent of their mother—even if they have never met. In one study, a researcher held a newly hatched zebra finch chick and puffed the scent of a parent near its nares. The chick's instinctive response to the puffs of air near its face was to beg for food, which chicks do by opening their mouths wide and waving their head around a bit. The researchers measured how long each chick begged to a series of differently scented air treatments, each with a different adult. In the first experiment, chicks were presented with the scent of a genetic parent (either mother or father) and the scent of an unrelated adult (the same sex as the parent). The chicks begged significantly longer in response to the scent of a genetic parent, compared to an unfamiliar adult. Then the researchers repeated the experiment, exchanging eggs between nests of unrelated female zebra finches and comparing the chicks' responses to the scent of the foster parents and genetic parents. The newly hatched chicks begged toward their genetic mother significantly longer than they did toward their foster mother, suggesting that they recognized, and were more stimulated by, the scent of their genetic mother, despite never having met.†

In fact, the olfactory powers of chicks begin to develop when they are still in the egg, prior to hatching. Embryonic chickens show behavioral responses to compounds known to occur in bird scent. Young nestlings recognize the odor of the nest they hatched in, even if they are removed from that nest soon after hatching, suggesting that they begin learning the scent of their nest even before they hatch.

---

* In barn swallow eggs, the female embryos give off higher concentrations of several volatile compounds compared to the male embryos, reflecting the same emphasis on female scent that we have noted in adult birds.

† Interestingly, they did not show the same discrimination toward their genetic father in comparison to their foster father.

Preen oil produced by female birds seems to matter for the health and survival of their eggs and nestlings, not only for its anti-microbial properties but also for the scent it gives off. It would be instructive to learn more about the sex differences, or lack thereof, in preen oil of species where both the males and females spend equal time on the nest.

WOMEN'S WORK

Examining the role of female birds in studies of avian chemical communication has led me to think about the role of the women who conduct that research. You may have noticed that there are many amazing women scientists contributing to this field, from its origins with Bernice Wenzel and Betsy Bang, to its inspiring progress through the work of Gabrielle Nevitt and Julie Hagelin, to the exploration of new frontiers with Barbara Caspers, Luisa Amo, Sarah Leclaire, and Leanne Grieves (just to name a few). It may seem like this field is dominated by women— indeed, many of the pioneers have been women—but I can easily name many men who have done research in this area, such as Jacques Balthazart, Francesco Bonadonna, Manuel Martín-Vivaldi, and Tobias Krause. I don't know for sure whether there are more women than men in avian chemical ecology, or even how that could be calculated. It may be one of those few scientific fields where men and women are about equally represented; though, due to ingrained prejudices, that very equality may lead to the field being perceived as mostly female.*

Scientists are often portrayed in movies and television as arrogant, highly competitive lone-wolf types who work in secret and fear being scooped by others. Women scientists, who make up only 18% of the scientists portrayed in film, are no exception—when

---

* Luca Turin notes in The Secret of Scent that although perfumery was historically an all-male profession, in recent years women have joined their ranks in equal numbers.

they do appear on screen, there is usually only one. Consider the paleobotanist Dr. Ellie Sattler in *Jurassic Park* and the linguist Dr. Louise Banks in *Arrival*. And as rare as their appearances are, it is even rarer for fictional women in science to be shown collaborating with each other.*

When I was growing up, I would proudly proclaim that I "wasn't like other girls." While I wasn't exactly a "tomboy," I was definitely not what we would've called a "girly-girl." I disdained anything conspicuously feminine, in fact, and I believed that boys made better friends than girls. Although ironically, when I reflect on that time, I see that most of my lasting friendships were actually with other girls. As an adult, I am an ardent feminist and strongly value the women in my life. But the internalized misogyny of my youth took many years to get past.

While I was in graduate school working on my PhD, I did not always see senior female academics as role models. I witnessed instances of women actively discouraging other women from getting involved in science, and I interpreted their behavior as an effect of viewing other women as threats. What I couldn't see at the time was that those women were sharing their negative experiences in an attempt to warn newcomers about how difficult it can be to be a woman in science, and how little support they received, in an effort to help prepare other women to face and overcome those challenges.

In retrospect, I realize that there were many examples of wonderful, supportive, successful women in my anthropology PhD program—students and faculty alike—who were committed and successful, yet still led full, joyful, and generous lives. But the stereotype of bitter, overly competitive academic women loomed so large in the popular imagination that I nearly believed it was the

---

* A notable exception is the 2018 film *Annihilation*, in which the five main characters are all women and scientists—but their collaboration doesn't exactly go well for any of them.

truth. I despaired of finding my way in science, and I wondered if it was even the right place for me.

Joining Dr. Ellen Ketterson's lab for my postdoc was a revelation. She was, and still is, one of the most important mentors in my life, and I think everyone who passed through or even near her lab felt the same way. She is inclusive and encouraging to others, and she is a highly successful scientist. Engaging with her style of doing science opened my eyes to a new way of interacting with the women around me, and it helped me to become the collaborative person I am today.

I met Dr. Julie Hagelin when I was first getting interested in studying avian olfaction. I was excited and a little intimidated about meeting her—she was one of the first people to study scent in birds, and her work on crested auklets was so interesting! But far from being standoffish, she was incredibly friendly and happy that I was interested in joining our burgeoning field. Like me, she got her start in the field as a postdoc, and she was excited about bringing together multiple levels of biological complexity, from olfactory cells in the nose, to complex processing in the brain, to behavioral responses of a bird detecting an odor. She encouraged me to attend my first scientific conference focused on odors—the annual meeting of AChemS, or the Association for Chemoreception Sciences. She claimed that whenever she went to this meeting, by the end of each day, her brain "ached" (in a good way) from taking in so much new, genuinely interdisciplinary information that challenged her to think differently. Julie introduced me to other scientists (including the avian olfaction pioneer Bernice Wenzel) and made sure people noticed my work. I consider her a true friend. We were excited to finally publish a paper together in 2021 and hope to collaborate more in the future.

My first direct interaction with Dr. Barbara Caspers was a bit of a surprise. In 2013, I had just published my paper showing a relationship between junco preen oil volatiles and the number of offspring those juncos produce. Out of the blue, I received an email from Caspers in which she apologized for a negative review that she had written of the manuscript when I had first submitted it. She felt that she had been too harsh. Note that she reached out this way, even knowing that her name would never be connected with her comments; the editors of most journals do not allow the identities of reviewers to be shared with the authors. This anonymity allows reviewers to be honest without fear of reprisal. I didn't remember any unduly punitive reviews of this particular paper, but I went back and searched my records to see if any stood out as particularly negative. I was baffled—although there were certainly some critical reviews, they had been constructive and helped me to improve the paper. I had no idea which one was hers, and I never asked. But her incredibly kind gesture stuck with me. I was touched, and I recommitted myself to being constructive and supportive of other scientists. Years later, I visited Barbara at the University of Bielefeld in Germany, where she is a professor. She has an amazing laboratory and a huge colony of zebra finches. To my delight, we've discussed collaborating on a project with her birds.

Many of my own collaborators in avian chemical ecology research are men. But, to me, there is something special about being part of a scientific subfield that is so heavily influenced by the work of women. I can't help but wonder, though, if one of the reasons that the existence of chemical communication in birds has been overlooked for so long is that many of the lead people researching it, especially early on, have been women. According to Julie Hagelin, after decades of pioneering contributions, Bernice Wenzel sometimes shook her head and remarked how she had felt like a "lone voice in the wilderness" within the ornithological community, which was strongly influenced by the "old boys' club" of leading ornithologists who were certain they already knew how birds perceived the world.

Without diversity in the people who do science, monolithic perspectives can blind us. The fact that women's discoveries have taken the lead in overturning the long-standing myth of avian anosmia—a myth perpetuated primarily by white men—clearly illustrates this principle. Women have also pointed out phenomena in female animal biology that have been long overlooked due to biased beliefs about behavior of the sexes. I am thrilled to see so many women in this field, and I have observed the welcoming effect they are having on the next generation of women scientists. The recent attention given by universities to increasing diversity, equity, and inclusion efforts is a promising sign, and it makes me hopeful that we can also broaden racial and ethnic representation in the biological sciences—I am painfully aware, for example, that of the many scientists I have mentioned by name in this book, only one is Black. It's quite possible that the problem is my own failure to seek out and highlight the work of researchers who are Black, Indigenous, or people of color. But I suspect that my field simply isn't terribly welcoming to people from underrepresented groups. Whether this exclusion is conscious or not, the result is that people are driven away from research. To see (or hear, or smell) the world as it truly is requires many different viewpoints. Without diverse voices and experiences, our knowledge is—and will remain—deficient.

A BREATH OF

FRESH AIR

"It makes everyone nervous, smelling," he says re the vial,
"because smell is such a strong sense." Turin gives talks on
smell to scientific audiences, and the squeamish reaction
pisses him off . . . "Real men and scientists feel slightly ridic-
ulous smelling something. I'll say 'Let me show you some
smells,' and I start passing out vials and everyone titters, like
I've just asked them to take off their clothes or something."
   —LUCA TURIN, quoted in *The Emperor of Scent:*
   A True Story of Perfume and Obsession by Chandler Burr

In August 2019, I attended a meeting at the University of Southern
California, and afterward I had a whole afternoon and evening to
myself since my flight back to Michigan didn't leave until the next
morning. Happily, the Natural History Museum of Los Angeles
County was just down the street. I love natural history museums.

   Once I finished touring that summer's excellent special exhibit on Antarctic vertebrate paleontology, I went upstairs to the
Jane G. Pisano Dinosaur Hall, where the museum houses its regular dinosaur collections. I noticed they had a section specifically

focused on dinosaur senses. I steeled myself against disappointment as I approached the area that discussed smell. However, I
was delighted to find a display noting that smell was likely impor_
tant to many dinosaurs, and to my astonishment, it even acknowledged that their modern descendants, birds, also use their sense
of smell.

Suddenly excited, I went to the Hall of Birds and found a display
on bird senses. The sign labeled "Smell" read, "The importance
of the sense of smell in birds has long been debated. Dr. Kenneth
Stager, former Curator of Birds and Mammals here at the Natural
History Museum, showed that Turkey Vultures use smell to locate
dead animals to eat. A sense of smell is important for other birds
as well. Ocean-wandering birds such as albatrosses, petrels, and
shearwaters use smell to find food, nest sites, and perhaps even
mates." I had forgotten that Stager, who established that turkey
vultures could detect the scent of the ethyl mercaptan added to natural gas, had once worked in that very Hall!

It's hard to describe the odd sense of relief I felt at seeing these
words. For years, I had known these scientific facts to be true, yet I
regularly encountered surprise, or even outright disbelief in scientists and nonscientists alike when I told them about my research.
The fact that birds have a robust sense of smell is still not common
knowledge.

But by the time I experienced this rare rush of vindication, my
work was starting to get noticed. A few months later, the Journal
of Experimental Biology published the study in which Kevin Theis
and I administered antibiotics to junco uropygial glands, changing the microbial community there and ultimately the odors that
they produced. A reporter from the New York Times interviewed
me and covered the work for their Science section, which is one
of the most exciting things that has ever happened to me.

I occasionally get emails from strangers, sharing their observations of bird scent and anecdotes about birds appearing to use

their sense of smell. Recently, I heard from a photographer in Ontario who had noticed that whenever they put out suet in a semi-
wild park area, woodpeckers would show up within minutes. The
photographer had come to the conclusion that it must be the smell
that attracted the birds and wanted to share this revelation with
me (as well as a beautiful photograph of a pileated woodpecker).

Birdwatchers, or birders, seem to be particularly interested in
hearing about my work. Birding is simultaneously highly acces-
sible and challenging—if you can get outdoors, you will almost certainly see some birds, but observing the less common, shyer species
will take time and patience. I suffer from a notable lack of patience,
and therefore I'm not a birder myself. In the early months of the
COVID-19 pandemic, like many other people sheltering at home,
I thought maybe I'd give birding a try, since I had nothing better
to do. For a while, I kept a list of the birds I saw on runs around my
neighborhood and on local trails, but I got bored after a few weeks.
I'm terrible at recognizing bird song, which really inhibits my ability to identify the birds around me. To me, birds are best when I can
hold them, see them up close, feel their feathers and their surprisingly strong bodies, and, yes, smell them. But I enjoy hearing about
how much pleasure birders get out of spotting new species and
behaviors, and I love to hear from them—this group of people is
interested in the world around them and is open to learning new
things.

That said, not all the emails I get are nice. Shortly after the New
York Times article appeared, I received a missive that read, "The
recent summary I read on your work concerns itself with various
chemicals and transformations associated with birds. It appears to
postulate that various 'perfumes' and/or odors are acting to influence the behavior of the birds. It would appear that it is taken for
granted that the birds 'smell' these chemicals, however I am not
aware of any definitive study that demonstrates that birds possess

olfactory systems akin to mammals, for example. Could you help me out by citing any reliable studies that prove birds smell? In the absence of such proof, it would seem that no definitive link exists between the chemicals and microbes you are studying, and their influence on birds' behaviors."

This impressive feat of mansplaining took me back over a decade to the day Jim Goodson told me so matter-of-factly that birds couldn't smell. The sender apparently found it necessary to also sign off with his credentials, specifying that he had received a PhD from a prestigious university three decades earlier. I debated whether to respond. Although he had gone to the trouble of finding my email address, he had obviously not read any of my papers or those that I had cited in my work. I couldn't figure out how to answer without making a snide quip about how to use Google, so I decided it was better to leave it alone.

I was frustrated by that email, and by the continued perpetuation of the myth. In the grand scheme of things, believing that birds do not have a sense of smell seems relatively harmless. But in this age of viral misinformation, it's increasingly important for students to learn how to find reliable information and to question unsubstantiated claims—and for them to continue using those skills as adults.

Jim Goodson, I must point out, good-naturedly accepted the results of my work and the news that he had been wrong about birds. He stopped teaching his students that birds don't use their sense of smell. I don't know whether he stopped throwing away the tiny olfactory bulbs from his brain dissections, but I hope so. He passed away from cancer in 2014, far too early, at the age of 48. I will be forever grateful to him for the unexpected way he nudged my research forward.

The science of smell is an unusual field in many ways. It's difficult to study because we can't see the molecules, or understand how another being senses them, or even accurately describe what we smell without highly specialized training. Perfumers and sommeliers have amazing olfactory skills, but there is a tendency to think they are "making it up" in order to sell a product. As a society, we are deeply uncomfortable with the idea of smell. Americans in particular are obsessed with cleanliness: we fear body odor and we complain about coworkers stinking up the office kitchen with their leftovers. Natural smells are shameful to us.

People also believe that smell isn't really that important to us. When asked which sense they would give up if they had to lose one, many people would choose smell because it seems so much less important in everyday life than sight or hearing. However, the pandemic that ravaged the world in 2020 and beyond has brought attention to the significance of smell, since the loss of it is a common symptom of COVID-19 infection. Some victims still suffer from the inability to detect scent long after they have recovered from the disease. Unfortunately, loss of olfactory capabilities is often linked to anxiety and depression and feelings of social isolation. This increase in the incidence of anosmia has highlighted the role that smell plays in our mental health and emotional well-being; perhaps people will be less quick to take this sense for granted from now on.

Science doesn't follow a straight line. Sometimes, new findings show that past studies were wrong, or incomplete, or missing important nuances. Often, scientific results raise more questions than they answer. They certainly have for me. For people seeking clear explanations, this can be disheartening. For all its shortcomings and inefficiencies, though, the scientific method is still the best means of inquiry available to us.

Life doesn't follow a straight line either. I can't imagine any way I could have planned my life so that I would end up where I am now. And I can't quite sense where I'm headed in the future. But I expect that I'll learn a lot of interesting things along the way.

Graduate students and others looking for career paths outside the usual tenure-track faculty route sometimes ask for my advice. I've never really known what to tell them, other than this: Say yes to opportunities that interest you. Don't focus on your dissertation or postdoctoral research to the exclusion of everything else— either personally or professionally.

When I took the job managing Ellen Ketterson's lab, some well-meaning professors in my graduate program suggested that those administrative responsibilities would hold me back from research achievements. But without that experience, I wouldn't have been qualified for my current position as managing director of a massive research center. Roller derby might seem like an unnecessary distraction to some people, but without it, I never would have discovered my mighty capacity for physical and mental endurance, both of which have served me well in my long days at the lab and in the field. Explore possibilities that come your way, even if they are unexpected. You never know what new revelations you'll have.

I urge everyone reading these words to never underestimate the power of basic science, or of learning for its own sake. Sometimes, when I tell nonscientists about my research, they cannot understand why on earth I would spend my time, and the government's money, on what they see as such a dumb thing. Who cares if birds can smell? But the implications are so much bigger than that. This simple question—do birds have a sense of smell?—has led me on a surprising journey into not just beaks, bulbs, and behaviors but also immune systems, hormones, genes, and bacteria. I suspect all good scientists find themselves following unanticipated paths deeper into the mysteries of life and the universe around us. Experiments don't always go as planned, and our ideas are not always supported. Still, we must pay attention to what our science tells us and have the curiosity and tenacity to follow our noses wherever they lead.

ACKNOWLEDGMENTS

. . . .

Authors always thank their editors in the acknowledgments, and now I fully understand why. This book wouldn't exist without my editor, Tiffany Gasbarrini, who decided to seek out opportunities to increase the visibility of women scientists. I was stunned to receive her email asking if I'd be interested in writing a book, and I am so glad I answered. Thank you to Tiffany and her amazing editorial assistant, Esther Rodriguez, for all their hard work turning my disorganized ramblings into an actual book. Also, many thanks to Kim Johnson and Juliana McCarthy for their diligent care of my manuscript, Amanda Weiss and Martha Sewall for the beautiful design, and all the staff at JHUP for their many efforts.

I am grateful to the many scientists who have mentored me and collaborated with me. Thank you, first and foremost, to Ellen Ketterson, who showed me that you can be a good scientist and a good person at a time in my life when I wasn't sure it was possible to be both. Thank you to Julie Hagelin for welcoming me into the world of bird smell, and for your comments on this book. My work would not be possible without interdisciplinary collaborators, and I am especially grateful to Helena Soini and Milos Novotny for their work analyzing the volatile chemical compounds in preen oil and to Kevin Theis for his work on junco microbiomes.

Thank you to my BEACON colleagues Charles Ofria, Erik Goodman, Louise Mead, Judi Brown Clarke, Connie James, Kay Holekamp, Rich Lenski, Rob Pennock, Tom Getty, Percy Pierre, Joseph Graves Jr., James Foster, Risto Miikkulainen, and Ben Kerr for your enthusiastic support of my research. Thank you to Joel Slade and Travis Hagey for taking a chance on a nontraditional academic as your postdoc adviser.

Thank you to the Junco Crew, especially fellow Ketterson alumni Jonathan Atwell, Christy Bergeon Burns, Kristal Cain, Nicole Gerlach, Jodie Jawor, Ryan Kiley, Abby Kimmitt, Dawn O'Neal, Mark Peterson, Dustin Reichard, Lynn Siefferman, Samuel Slowinski, Eric Snajdr, and Sarah Wanamaker. Extra special thanks to Kim Rosvall, who for years has enthusiastically informed me, "You're revolutionizing the field!"

Much of this book was written or edited at Community-Engaged Scholarship writing events at Michigan State University. Thank you to Diane Doberneck at MSU's University Outreach and Engagement, who organized these events and offered enthusiastic encouragement for my project.

Thank you to the League of Extraordinary Officials, my close-knit group of roller derby friends who have cheered me on, sometimes literally: Laurie "Bibbity Bobbity Boom" Van Egeren, Cyndi "geoknitter" Rachol, Rosangela "Bert Hert" Canino-Koning, Kendall "Godwin's Lawyer" Koning, Alexander "Technical Difficulty" Haase, Leigh "Ph.Demon" VanHandel, Daniel "Adam Smasher" Alt, Laura "Wishbone Breaker" Carr, Scott "Scotchy Scotch Scott" Carlson, Greg "Dr. Math" Muller, Chris "Dread Hochuli" Scheper, Tiffany "Chemical Restraint" Kasper, Allie "Tornado Allie" Schultze, Wendy "Meow KaPow" Whitlock, and Bethany "Biz Barksalot" Adams. Special thanks to Nicole "Smaxx Kittenz" Cottom for my author photo.

Thank you to my Best Friend Forever, Angela Malcolm Stucker, for your unconditional support. I always know that you're there for me, even when our introverted tendencies and hectic lives mean we sometimes don't see each other or talk for months on end. And finally, thank you to my husband, Nathan Burroughs, who puts up with my endless travels and listens to my weird ideas, who convinced me I could write this book and was my first reader, and who has truly helped me live my best life.

GLOSSARY

· · · · · · ·

ALLELES. Different versions of the same gene (from the Greek word for "other"). For example, in Gregor Mendel's famous experiments with pea plants, he discovered that the plants have two versions of the gene that controls flower color: one purple, the other white.

ANOSMIA. The absence or loss of smell. From the Greek *osmē*, odor.

ANTIGEN. A small peptide from an invasive agent, such as a virus or bacterium, that stimulates an immune response in the host.

BILL-WIPING. A behavior in which a bird wipes its bill on a branch or other substrate, performed to clean the bill or during social interactions. This behavior was previously believed to be an irrelevant, nervous behavior when performed in social situations. Recently, I suggested that it may function as an olfactory display, because it is common during courtship and may release odors trapped in residual preen oil on the bill.

B LYMPHOCYTES (SOMETIMES JUST CALLED "B CELLS"). White blood cells that secrete antibodies to fight pathogenic bacteria and viruses.

BROODING. The act of sitting on top of nestlings in the nest to keep them warm (similar to incubation), usually performed by a parent bird.

BROOD PATCH. A featherless, vascularized area on a bird's chest that increases skin contact to eggs and nestlings to keep them warm.

CHEMICAL CUE. Any compound or blend of compounds that contains information used by a receiver. In contrast to a chemical signal, a cue has not necessarily evolved for this purpose.

CHEMICAL ECOLOGY. The study of chemically mediated interactions between organisms, including via smell, and how these interactions influence the behavior and evolution of those organisms.

CHEMICAL SIGNAL. Information-containing compound or blend of compounds that has evolved through natural selection because of its benefits for both the sender and the receiver (in contrast to a chemical cue).

CHROMOSOME. A long strand of DNA, which coils up into a tight structure during cell division. Individuals receive one copy of each chromosome from each parent, resulting in paired chromosomes. Humans have twenty-three pairs of chromosomes.

CLOACA. The posterior orifice that serves as the opening for the digestive, urinary, and reproductive tracts in vertebrate animals except for most mammals. Amphibians, reptiles, birds, monotremes (egg-laying mammals), and many fish all have a cloaca. The word comes from the Latin *cluere*, to cleanse, and can also mean "sewer."

CLUTCH. A group of eggs fertilized at the same time, laid in a single session (usually stretching over days) and incubated at the same time. The average number of eggs per clutch varies by species. Dark-eyed juncos typically have about four eggs in a clutch.

COMMON GARDEN EXPERIMENT. A method in which researchers take young individuals from different environments and raise them in identical, controlled conditions, allowing the effects of genetics and environment to be examined separately.

CONSPECIFIC. A member of the same species.

COOPERATIVE BREEDING. A mating system in which a socially and/or genetically monogamous mated pair receives help raising their offspring from adult, nonreproductive individuals in their group.

COPROPHAGY. From the Greek *copros*, "feces," and *phagein*, "to eat," which should tell you all you need to know.

CORTICOSTERONE. A steroid hormone secreted by the adrenal gland and involved in the stress response. Corticosterone is the primary hormone serving this role in amphibians, reptiles, birds, and rodents. The analog in most mammals, including humans, is cortisol.

CROSS-FOSTERING. A method in which researchers switch the same-aged offspring of two different sets of parents, allowing the effects of genetics and environment to be examined separately.

CROSSING-OVER. The exchange of genetic material between paired chromosomes during meiosis, after DNA replication but before cell division. This process results in a chromosome that contains material from two different parents.

CUE. Any trait that communicates information about the trait-holder to another individual. A cue can exist in any sensory modality (e.g., chemical, visual, audio). In contrast to a signal, a cue has not necessarily evolved for this purpose.

DIMETHYL SULFIDE (DMS). A small volatile compound given off by patches of phytoplankton in the ocean as a by-product of their metabolic processes. Known to attract foraging seabirds, high concentrations of this odor indicate the presence of fish, squid, and other animals preyed upon by these birds.

DISASSORTATIVE MATING. Choosing a mate that is different from oneself at a particular trait, such as MHC genotype.

DNA SEQUENCING. A method used to determine the sequence of nucleotide bases (A, C, G, or T) in a stretch of DNA. These techniques take advantage of the fact that DNA copies itself under certain conditions, and they provide fluorescently labeled nucleotides (the "letters" in a strand of DNA) in a chemical mixture. The sequencing machine then detects these tagged molecules and can use this information to "read" the order, or sequence, of the DNA.

ESTER. An organic compound that results when a hydrogen atom from an acid is replaced by an alkyl group. For example, many naturally occurring fats and essential oils are esters. Preen oil can include monoester waxes, which only have one alkyl group, or heavier diester waxes, which have two alkyl groups, among other compounds.

EXTRA-PAIR COPULATIONS/FERTILIZATIONS. Mating events that occur between two birds that are not members of the same bonded pair.

EXTRA-PAIR OFFSPRING/YOUNG. Offspring that result from extra-pair fertilizations, usually revealed through paternity testing.

GAS CHROMATOGRAPHY–MASS SPECTROMETRY (GC-MS). An analytical chemistry technique used to identify chemical compounds in a mixture. The method separates different compounds in a gas on the basis of mass and electric charge, and it measures how much of each compound is in the mixture.

GENE. A stretch of DNA that does something. Functional genes perform biological tasks, such as protein coding (the sequence of bases indicates the sequence of amino acids to be linked together) or regulating the functions of other genes. In contrast, other stretches of DNA do nothing, like microsatellites or pseudogenes.

GENE DUPLICATION. A mutational process by which a stretch of DNA is accidentally included twice when an individual's gametes (eggs or sperm) are formed during meiosis; this extra DNA is passed on to the offspring.

GENETIC MONOGAMY. A mating system in which a male and a female exclusively copulate with each other.

GENOTYPE. A description of the alleles carried by an individual at a particular locus. For example, a pea plant could have the genotype of two white flower alleles, two purple flower alleles, or one of each.

GERM-FREE ORGANISMS. Organisms (often lab mice or rats) developed in a sterile environment and lacking any bacteria or other microorganisms.

HETEROSPECIFIC. A member of a different species.

HETEROZYGOUS. Having two different alleles for the same gene (one inherited from each parent).

HISTOCOMPATIBILITY. Tissue compatibility resulting from a tissue donor and a tissue recipient having similar enough genes that the recipient's immune system does not reject the donated tissue.

HOLOBIONT. A multicellular host (such as an animal or a plant) and its symbiotic microbiome considered as a single entity. From the Greek *hólos*, "whole," and *biont*, an English suffix used in biology to refer to a living organism.

HOLOGENOME. The collection of all of the genes present in the holobiont.

HOMOZYGOUS. Having two identical alleles for the same gene (one inherited from each parent).

HORMONE ASSAYS. A method to measure how much of a hormone is present in blood or another type of biological sample

IMMUNOHISTOCHEMISTRY. A technique for visualizing components of tissue under a microscope. It selectively identifies proteins in the cells of a tissue section by staining the tissue with dyes or enzymes that only react with specific protein targets. As a result, only the targeted proteins appear colored in the image, making them easily visible under the microscope.

INCUBATION. The act of sitting on top of eggs in the nest to keep them warm (similar to brooding) and help the embryos develop, usually performed by a parent bird.

LEK. A mating system in which males gather together to display all at once, competing for the attention of females. Females choose which male to mate with based on these displays. After mating, the males usually have no involvement in raising the offspring. From the Swedish *lek*, "play."

LIGAND. A peptide that forms a bond with another peptide, often producing a signal to other cells as a result. From the Latin *ligare*, "to bind."

LINKED GENES. Genes that are located near each other on a chromosome, reducing the likelihood that they will be separated during meiotic crossing-over events and increasing the probability that they will be inherited together over generations.

LOCUS, *plural* **LOCI.** From the Latin word for "place," referring to the location of a gene or other marker on a strand of DNA.

MACROPHAGE. A white blood cell that engulfs and digests particles and cells that are deemed outsiders. From the Greek for "large eaters."

MACROSMATIC. Having a strongly developed sense of smell.

MAJOR HISTOCOMPATIBILITY COMPLEX (MHC). A large family of immune genes that allow the immune system to recognize when invasive pathogens are present. In humans, MHC is called "human leukocyte antigen," or HLA.

MEIOSIS. A type of cell division that produces germ cells (which then become gametes such as eggs and sperm) that only have one copy of each chromosome. Meiosis begins with DNA replication, in which the entire genome is copied, followed by two rounds of cell division.

MICROBIOME. The collection of microorganisms in a particular environment, such as the gut microbiome or the uropygial gland microbiome.

MICROSATELLITES. Repeated small stretches of DNA. Because they are nonfunctional and do not produce any proteins or regulate any other genes, they are free to mutate more often than functional genes, resulting in high variation across individuals. Microsatellites are commonly used in paternity tests and as a proxy for measuring an individual's overall genetic diversity.

MICROSMATIC. Having a poorly developed sense of smell.

MIST NETS. These nets are made of fine-mesh nylon or polyester panels and are used to catch birds or bats. The nets have horizontal lines that create large, loose pockets. They are called "mist nets" because they are very difficult to see, so the animals fly into them and become quickly entangled. They are generally very safe for the animals when used by an experienced handler.

NARES. The bird analog of nostrils, the nares are two small holes on the top of the beak, close to the face. The singular form of the word is naris, from the Latin *naris*, "nostril" or "nose."

NEXT-GENERATION SEQUENCING. In contrast to traditional Sanger DNA sequencing methods, next-generation methods are able to sequence millions of short pieces of DNA simultaneously.

OLFACTION. The sense of smell, or the action of smelling.

OLFACTORY BULB. A neural structure located near the front of the brain which receives information from the olfactory receptors and passes it along to other brain areas (including the amygdala and the hippocampus, involved in emotion, memory, and learning) for further processing.

OLFACTORY RECEPTOR. A protein expressed in olfactory receptor neurons, located in the lining of the nasal passageways, that binds to odorant molecules. Animals typically have hundreds of different olfactory receptors, allowing the animal to detect a wide range of odors. When the receptor protein binds to an odorant molecule, it triggers a response in the neuron, sending information to the olfactory bulb.

PEPTIDE. A fragment of a protein.

PHENOTYPE. An expressed trait, which can be influenced both by the individual's genotype and the environment in which they developed.

PHEROMONE. A semiochemical that evokes a specific reaction in the recipient.

PHILOPATRY. A social system in which members of one sex stay in the groups in which they were born, common in animals with large social groups, such as hyenas and baboons. From the Greek philos, "beloved," and patra, "homeland."

POLYGYNY. A mating system in which one male lives in a group with and mates with many females. From the Greek poly, "many," and gyne, "wife."

POLYMERASE CHAIN REACTION (PCR). A technique in which small stretches of DNA are copied or "amplified" by exposing the original DNA and other reagents to repeated cycles of heating and cooling, which stimulates the DNA strand to copy itself. Biochemist Kary Mullis was awarded the Nobel Prize in Chemistry for the invention of this technique, and claimed that LSD helped him develop the concept.

PREEN OIL. A fatty, waxy substance secreted by the uropygial gland. Birds use their bills to take this oil from the gland and spread it onto their feathers, where it helps to maintain the feathers' integrity and appearance and to protect the birds from parasites. Preen oil also contains volatile compounds that contribute to a bird's body odor.

PSEUDOGENES. Genes that were once functional in the evolutionary past but no longer produce the proper protein sequences due to mutations.

ROLLER DERBY. A full contact sport in which players wear skates and compete on an oval track to score points by passing each other. Modern flat track roller derby was created in Austin, Texas, in 2003.

RUFF SNIFF. A behavior in which a bird buries its bill in the nape feathers of a social or courtship partner. Commonly observed in crested auklets and believed to function as a way to smell the other bird's strong tangerine-like odor, which is only produced during breeding season.

SCENT-MARKING. A behavior in which an animal deposits urine, feces, or an odiferous secretion from a scent gland on a surface in its environment.

SEMIOCHEMICAL. Any kind of compound involved in the chemical interaction between organisms. From the Greek *semeion*, "mark" or "signal."

SEXUAL DIMORPHISM. The genetically inherited condition wherein sexes of the same species exhibit different physical characteristics. One example is the northern cardinal; females appear mainly gray, while males are bright red.

SIGNAL. Information-containing trait that has evolved through natural selection because of its benefits in the context of communication for both the sender and the receiver (in contrast to a cue).

SIGNATURE MIXTURES. Semiochemicals that contain information about the individual sender.

SIMULATED TERRITORIAL INTRUSION (STI). A behavioral trial in which a researcher uses one or more stimuli (a bird in a cage, a playback of bird song, etc.) on a wild animal's territory to elicit a response. Responses of the territory-holder can range from aggression to courtship.

SOCIAL MONOGAMY. A mating system in which the male and female are pair bonded, spending the majority of their time together, at least for the length of one breeding season. Socially monogamous pairs may or may not also practice genetic monogamy.

TESTOSTERONE. A steroid hormone that is typically present in higher concentrations in males than females and stimulates the development of male sexual characteristics. Testosterone is produced in the testes, as well as in the ovaries and the adrenal cortex.

UROPYGIAL GLAND. A sebaceous gland located above the base of the tail in nearly all bird species. Also called the preen gland, it secretes preen oil, produced in the two lobes of the gland under the skin and secreted through the papilla that can be observed on the surface of the skin. From the Ancient Greek *oura*, "tail," and *puge*, "rump."

VECTOR. An organism, often an arthropod, that carries and transmits an infectious agent (such as a virus or a parasite) from an infected organism to another organism. Mosquitoes are a vector for malaria.

VOLATILE COMPOUNDS. Small chemical compounds that have a tendency to vaporize at room temperature. Larger compounds evaporate more slowly and may be referred to as semivolatile.

VOMERONASAL ORGAN. An accessory olfactory organ located in the nasal cavity, thought to be important in detecting pheromones and other nonvolatile organic compounds. Found in many vertebrates but not birds.

WITHIN-PAIR OFFSPRING/YOUNG. Offspring that result from copulations between a mated pair.

Y-MAZE. An apparatus used to test behavioral preference or avoidance when a test animal is given two choices. It is shaped like a Y, with a neutral area at the base of the Y, where the test animal is placed, and the two choices located at the end of the Y arms, one in each arm.

REFERENCES

....

CHAPTER 1. THE MOST ANCIENT AND FUNDAMENTAL SENSE

Wenzel, B. M. 1967. "Olfactory perception in birds." In *Olfaction and Taste II: Proceedings of the Second International Symposium*, edited by T. Hayashi, 203–218. London: Pergamon Press.

Barrowclough, G. F., J. Cracraft, J. Klicka, and R. M. Zink. 2016. "How many kinds of birds are there and why does it matter?" *PLOS One* 11:e0166307.

Tedore, C., and D.-E. Nilsson. 2019. "Avian UV vision enhances leaf surface contrasts in forest environments." *Nature Communications* 10:238. https://doi.org /10.1038/s41467-018-08142-5.

Bang, B. G., and S. Cobb. 1968. "The size of the olfactory bulb in 108 species of birds." *The Auk* 85 (1):55–61.

Olkowicz, S., M. Kocourek, R. Lučan, M. Porteš, W. T. Fitch, S. Herculano-Houzel, and P. Nemec. 2016. "Birds have primate-like numbers of neurons in the forebrain." *Proceedings of the National Academy of Sciences* 113 (26):7255–7260.

Kirsch, J., O. Güntürkün, and J. Rose. 2008. "Insight without cortex: Lessons from the avian brain." *Consciousness and Cognition* 17 (2):475–483.

Jarvis, E. D., O. Güntürkün, L. Bruce, A. Csillag, H. Karten, W. Kuenzel, L. Medina et al. 2005. "Avian brains and a new understanding of vertebrate brain evolution." *Nature Reviews Neuroscience* 6 (2):151–159. https://doi.org/10.1038 /nrn1606.

Audubon, J. J. 1826. "Account of the habits of the turkey buzzard (*Vultur aura*) particularly with the view of exploding the opinion generally entertained of its extraordinary power of smelling." *Edinburgh New Philosophical Journal* 2:172–184.

Waterton, C. 1837. *Essays on Natural History, Chiefly Ornithology*. London: Longman, Brown, Green, and Longmans, quotes at pp. 34, 35.

McCartney, W. 1968. "Wheresoever the carcase is." In *Olfaction and Odours*. Berlin: Springer.

London Museum. n.d. "The New London Texas school explosion." London Museum, New London, Texas. Accessed September 18, 2019. http://www.nlsd.net/index2 .html.

Vine, K. 2007. "'Oh, my God! It's our children!'" *Texas Monthly*, March 2007. https:// www.texasmonthly.com/articles/oh-my-god-its-our-children/.

Stager, K. E. 1962. "The role of olfaction in food location by the turkey vulture (*Cathartes aura*)." PhD diss., Zoology, University of Southern California.

Wenzel, B. M. 1967. "Olfactory perception in birds." In *Olfaction and Taste II: Proceedings of the Second International Symposium*, edited by T. Hayashi, 203–218. London: Pergamon Press.

Wenzel, B. M. 1968. "Olfactory prowess of the kiwi." *Nature* 220:1133–1134.

Hutchison, L. V., and B. M. Wenzel. 1980. "Olfactory guidance in foraging by Procellariiforms." *The Condor* 82 (3):314–319.

Bang, B. G. 1960. "Anatomical evidence for olfactory function in some species of birds." *Nature* 188 (4750):547–549.

Bang, B. G., and S. Cobb. 1968. "The size of the olfactory bulb in 108 species of birds." *The Auk* 85 (1):55–61.

Bang, B. G., and B. M. Wenzel. 1985. "Nasal cavity and olfactory system." In *Form and Function in Birds*, edited by A. S. King and J. McLelland, 195–225. London: Academic Press.

McGann, J. P. 2017. "Poor human olfaction is a 19th-century myth." *Science* 356 (6338):eaam7263.

Negus, V. 1958. *Comparative Anatomy and Physiology of the Nose and Paranasal Sinuses*. Edinburgh: E. & S. Livingstone.

Porter, J., B. Craven, R. M. Khan, S.-J. Chang, I. Kang, B. Judkewitz, J. Volpe, G. Settles, and N. Sobel. 2007. "Mechanisms of scent-tracking in humans." *Nature Neuroscience* 10 (1):27–29.

Regenbogen, C., J. Axelsson, J. Lasselin, D. K. Porada, T. Sundelin, M. G. Peter, M. Lekander, et al. 2017. "Behavioral and neural correlates to multisensory detection of sick humans." *Proceedings of the National Academy of Sciences* 114 (24):6400–6405.

Trivedi, D. K., E. Sinclair, Y. Xu, D. Sarkar, C. Walton-Doyle, C. Liscio, P. Banks, et al. 2019. "Discovery of volatile biomarkers of Parkinson's disease from sebum." *ACS Central Science* 5 (4):599–606.

Frumin, I., O. Perl, Y. Endevelt-Shapira, A. Eisen, N. Eshel, I. Heller, M. Shemesh, et al. 2015. "A social chemosignaling function for human handshaking." *eLife* 4:e05154.

Miller, S. L., and J. K. Maner. 2009. "Scent of a woman: Men's testosterone responses to olfactory ovulation cues." *Psychological Science* 21 (2):276–283.

Gildersleeve, K. A., M. G. Haselton, C. M. Larson, and E. G. Pillsworth. 2012. "Body odor attractiveness as a cue of impending ovulation in women: Evidence from a study using hormone-confirmed ovulation." *Hormones and Behavior* 61 (2): 157–166.

Moran, D., R. Softley, and E. J. Warrant. 2015. "The energetic cost of vision and the evolution of eyeless Mexican cavefish." *Science Advances* 1 (8):e1500363.

Wyatt, T. D. 2014. *Pheromones and Animal Behavior: Chemical Signals and Signatures*. 2nd ed. Cambridge: Cambridge University Press.

Hölldobler, B., and N. F. Carlin. 1987. "Anonymity and specificity in the chemical communication signals of social insects." *Journal of Comparative Physiology A: Neuroethology, Sensory, Neural, and Behavioral Physiology* 161:567–581.

Maynard Smith, J., and D. Harper. 2003. *Animal Signals*. Oxford: Oxford University Press.

Laidre, M. E., and R. A. Johnstone. 2013. "Animal signals." *Current Biology* 23 (18):R829–R833.

Nevitt, G. A., R. R. Veit, and P. Kareiva. 1995. "Dimethyl sulphide as a foraging cue for Antarctic Procellariiform seabirds." *Nature* 376 (6542):680–682.

Nevitt, G. A. 2000. "Olfactory foraging by Antarctic Procellariiform seabirds: Life at high Reynolds numbers." *Biological Bulletin* 198:245–253.

Hagelin, J. C., I. L. Jones, and L. E. L. Rasmussen. 2003. "A tangerine-scented social odour in a monogamous seabird." *Proceedings of the Royal Society B: Biological Sciences* 270:1323–1329.

Hagelin, J. C. 2007. "The citrus-like scent of crested auklets: Reviewing the evidence for an avian olfactory ornament." *Journal of Ornithology* 148 (Supplement 2): S195–S201.

Bonadonna, F., and V. Bretagnolle. 2002. "Smelling home: A good solution for burrow-finding in nocturnal petrels?" *Journal of Experimental Biology* 205: 2519–2523.

Bonadonna, F., G. B. Cunningham, P. Jouventin, F. Hesters, and G. A. Nevitt. 2003. "Evidence for nest-odour recognition in two species of diving petrel." *Journal of Experimental Biology* 206:3719–3722.

Bonadonna, F., and G. A. Nevitt. 2004. "Partner-specific odor recognition in an Antarctic seabird." *Science* 306:835.

Roth, T. C., II, J. G. Cox, and S. L. Lima. 2008. "Can foraging birds assess predation risk by scent?" *Animal Behaviour* 76:2021–2027.

Amo, L., I. Galván, G. Tomás, and J. J. Sanz. 2008. "Predator odour recognition and avoidance in a songbird." *Functional Ecology* 22:289–293.

Steiger, S. S., A. E. Fidler, M. Valcu, and B. Kempenaers. 2008. "Avian olfactory receptor gene repertoires: Evidence for a well-developed sense of smell in birds?" *Proceedings of the Royal Society B: Biological Sciences* 275 (1649):2309–2317.

Clark, L., K. V. Avilova, and N. J. Bean. 1993. "Odor thresholds in passerines." *Comparative Biochemistry and Physiology Part A: Physiology* 104A (2):305–312.

CHAPTER 2. FOLLOWING THE BIRD'S NOSE

Whittaker, D. J., J. C. Morales, and D. J. Melnick. 2007. "Resolution of the Hylobates phylogeny: Congruence of mitochondrial D-loop sequences with molecular, behavioral, and morphological data sets." *Molecular Phylogenetics and Evolution* 45:620–628.

Whittaker, D. J. 2009. "Phylogeography of Kloss's gibbon (Hylobates klossii) populations and implications for conservation planning in the Mentawai Islands." In *The Gibbons: New Perspectives on Small Ape Socioecology and Population Biology*, edited by S. Lappan and D. J. Whittaker, 73–89. New York: Springer.

Whittaker, D. J. 2005. "New population estimates for the Kloss's gibbon (Hylobates klossii)." *Oryx* 39:458–461.

Wingfield, J. C., G. F. Ball, A. M. Dufty, R. E. Hegner, and M. Ramenofsky. 1987. "Testosterone and aggression in birds." *American Scientist* 75:602–608.

Wingfield, J. C., R. E. Hegner, A. M. Dufty, and G. F. Ball. 1990. "The 'challenge hypothesis': Theoretical implications for patterns of testosterone secretion, mating systems, and breeding strategies." *American Naturalist* 136:829–846.

Ketterson, E. D., V. Nolan Jr., J. M. Cawthorn, P. G. Parker, and C. Ziegenfus. 1996. "Phenotypic engineering: Using hormones to explore the mechanistic and functional bases of phenotypic variation in nature." *Ibis* 138 (1):70–86.

Ketterson, E. D., V. Nolan Jr., W. L. Wolf, and C. Ziegenfus. 1992. "Testosterone and avian life histories: Effects of experimentally elevated testosterone on behavior and correlates of fitness in the dark-eyed junco (*Junco hyemalis*)." *American Naturalist* 140:980–990.

Cawthorn, J. M., D. Morris, E. D. Ketterson, and V. Nolan Jr. 1998. "Influence of elevated testosterone on nest defence in dark-eyed juncos." *Animal Behaviour* 56:617–621.

Enstrom, D. E., E. D. Ketterson, and V. Nolan Jr. 1997. "Testosterone and mate choice in the dark-eyed junco." *Animal Behaviour* 54:1135–1146.

Chandler, C. R., E. D. Ketterson, and V. Nolan Jr. 1997. "Effects of testosterone on spatial activity in free-ranging male dark-eyed juncos when their mates are fertile." *Animal Behaviour* 54:543–549.

Raouf, S. A., P. G. Parker, E. D. Ketterson, V. Nolan Jr., and C. Ziegenfus. 1997. "Testosterone affects reproductive success by influencing extra-pair fertilizations in male dark-eyed juncos (Aves: *Junco hyemalis*)." *Proceedings of the Royal Society B: Biological Sciences* 264:1599–1603.

Casto, J. M., V. Nolan Jr., and E. D. Ketterson. 2001. "Steroid hormones and immune function: Experimental studies in wild and captive dark-eyed juncos (*Junco hyemalis*)." *American Naturalist* 157:408–420.

Nolan, V., Jr., E. D. Ketterson, C. Ziegenfus, C. R. Chandler, and D. P. Cullen. 1992. "Testosterone and avian life histories: Effects of experimentally elevated testosterone on molt and survival in male dark-eyed juncos." *The Condor* 94:364–370.

Ketterson, E. D., V. Nolan Jr., and M. Sandell. 2005. "Testosterone in females: Mediator of adaptive traits, constraint on the evolution of sexual dimorphism, or both?" *American Naturalist* 166:S85–S98.

Clotfelter, E. D., D. M. O'Neal, J. M. Gaudioso, J. M. Casto, I. M. Parker-Renga, E. Snajdr, D. L. Duffy, V. Nolan Jr., and E. D. Ketterson. 2004. "Consequences of elevating plasma testosterone in females of a socially monogamous songbird: Evidence of constraints on male evolution?" *Hormones and Behavior* 46:171–178.

Gerlach, N. M., and E. D. Ketterson. 2013. "Experimental elevation of testosterone lowers fitness in female dark-eyed juncos." *Hormones and Behavior* 63 (5):782–790.

O'Neal, D. M., D. G. Reichard, K. Pavlis, and E. D. Ketterson. 2008. "Experimentally-elevated testosterone, female parental care, and reproductive success in a songbird, the dark-eyed junco (*Junco hyemalis*)." *Hormones and Behavior* 54:571–578.

Zysling, D. A., T. J. Greives, C. W. Breuner, J. M. Casto, G. E. Demas, and E. D. Ketterson. 2006. "Behavioral and physiological responses to experimentally elevated testosterone in female dark-eyed juncos (*Junco hyemalis carolinensis*)." *Hormones and Behavior* 50:200–207.

Piersma, T., M. Dekker, and J. S. Sinninghe Damsté. 1999. "An avian equivalent of make-up?" *Ecology Letters* 2:201–203

Reneerkens, J., T. Piersma, and J. S. Sinninghe Damsté. 2005. "Switch to diester preen waxes may reduce avian nest predation by mammalian predators using olfactory cues." *Journal of Experimental Biology* 208:4199–4202.

Soini, H. A., S. E. Schrock, K. E. Bruce, D. Wiesler, E. D. Ketterson, and M. V. Novotny. 2007. "Seasonal variation in volatile compound profiles of preen gland secretions of the dark-eyed junco (*Junco hyemalis*)." *Journal of Chemical Ecology* 33 (1):183–198.

Whittaker, D. J., D. G. Reichard, A. L. Dapper, and E. D. Ketterson. 2009. "Behavioral responses of nesting female dark-eyed juncos *Junco hyemalis* to hetero- and conspecific passerine preen oils." *Journal of Avian Biology* 40:579–583.

CHAPTER 3. DECIPHERING THE SECRETS OF SMELLS

Mainland, J. D., A. Keller, Y. R. Li, C. Trimmer, L. L. Snyder, A. H. Moberly, K. A. Adipietro, et al. 2014. "The missense of smell: Functional variability in the human odorant receptor repertoire." *Nature Neuroscience* 17 (1):114–120.
Haberly, L. B. 1998. "Olfactory cortex." In *The Synaptic Organization of the Brain*, edited by G. M. Shepherd, 377–416. New York: Oxford University Press.
Majid, A., and N. Burenhult. 2014. "Odors are expressible in language, as long as you speak the right language." *Cognition* 130 (2):266–270.
Herz, R. S., and J. von Clef. 2001. "The influence of verbal labeling on the perception of odors: Evidence for olfactory illusions?" *Perception* 30:381–391.
Whittaker, D. J., H. A. Soini, J. W. Atwell, C. Hollars, M. V. Novotny, and E. D. Ketterson. 2010. "Songbird chemosignals: Volatile compounds in preen gland secretions vary among individuals, sexes, and populations." *Behavioral Ecology* 21:608–614.
De Groof, G., H. Gwinner, S. S. Steiger, B. Kempenaers, and A. Van der Linden. 2010. "Neural correlates of behavioural olfactory sensitivity changes seasonally in European starlings." *PLOS One* 5 (12):e14337.
Whittaker, D. J., H. A. Soini, N. M. Gerlach, A. L. Posto, M. V. Novotny, and E. D. Ketterson. 2011. "Role of testosterone in stimulating seasonal changes in a potential avian chemosignal." *Journal of Chemical Ecology* 37:1349–1357.
Ketterson, E. D., V. Nolan Jr., and M. Sandell. 2005. "Testosterone in females: Mediator of adaptive traits, constraint on the evolution of sexual dimorphism, or both?" *American Naturalist* 166:S85–S98.
Whittaker, D. J., N. M. Gerlach, S. P. Slowinski, K. P. Corcoran, A. D. Winters, H. A. Soini, M. V. Novotny, E. D. Ketterson, and K. R. Theis. 2016. "Social environment has a primary influence on the microbial and odor profiles of a chemically signaling songbird." *Frontiers in Ecology and Evolution* 4:90. https://doi.org/10.3389/fevo.2016.00090.
Shaw, C. L., J. E. Rutter, A. L. Austin, M. C. Garvin, and R. J. Whelan. 2011. "Volatile and semivolatile compounds in gray catbird uropygial secretions vary with age and between breeding and wintering grounds." *Journal of Chemical Ecology* 37 (4):329–339.
Karlsson, A.-C., P. Jensen, M. Elgland, K. Laur, T. Fyrner, P. Konradsson, and M. Laska. 2010. "Red junglefowl have individual body odors." *Journal of Experimental Biology* 213:1619–1624.
Célérier, A., C. Bon, A. Malapert, P. Palmas, and F. Bonadonna. 2011. "Chemical kin label in seabirds." *Biology Letters* 7 (6):807–810.
Kavaliers, M., E. Choleris, and D. W. Pfaff. 2005. "Genes, odours and the recognition of parasitized individuals by rodents." *Trends in Parasitology* 21 (9):423–429.
de Boer, J. G., A. Robinson, S. J. Powers, S. L. G. E. Burgers, J. C. Caulfield, M. A. Birkett, R. C. Smallegange, et al. 2017. "Odors of *Plasmodium falciparum*–infected participants influence mosquito-host interactions." *Scientific Reports* 7:9283.
De Moraes, C. M., N. M. Stanczyk, H. S. Betz, H. Pulido, D. G. Sim, A. F. Read, and M. C. Mescher. 2014. "Malaria-induced changes in host odors enhance

mosquito attraction." *Proceedings of the National Academy of Sciences* 111 (30): 11079–11084.

World Health Organization. 2018. "This year's world malaria report at a glance." https://www.who.int/malaria/media/world-malaria-report-2018/en/.

Grieves, L. A., T. R. Kelly, M. A. Bernards, and E. A. Macdougall-Shackleton. 2018. "Malarial infection alters wax ester composition of preen oil in songbirds: Results of an experimental study." *The Auk* 135 (3):767–776.

Grieves, L. A., and E. A. Macdougall-Shackleton. 2020. "No evidence that songbirds use odour cues to avoid malaria-infected conspecifics." *Behaviour* 157 (8–9):835–853.

Díez-Fernández, A., J. Martínez-de la Puente, L. Gangoso, P. López, R. Soriguer, J. Martín, and J. Figuerola. 2020. "Mosquitoes are attracted by the odour of *Plasmodium*-infected birds." *International Journal for Parasitology* 50:569–575.

Leclaire, S., W. F. D. van Dongen, S. Voccia, T. Merkling, C. Ducamp, S. A. Hatch, P. Blanchard, E. Danchin, and R. H. Wagner. 2014. "Preen secretions encode information on MHC similarity in certain sex-dyads in a monogamous seabird." *Scientific Reports* 4:6920. https://doi.org/10.1038/srep06920.

Slade, J. W. G., M. J. Watson, T. R. Kelly, G. B. Gloor, M. A. Bernards, and E. A. Macdougall-Shackleton. 2016. "Chemical composition of preen wax reflects major histocompatibility complex similarity in songbirds." *Proceedings of the Royal Society B: Biological Sciences* 283:20161966. https://doi.org/10.1098/rspb.2016.1966.

Grieves, L. A., G. B. Gloor, M. A. Bernards, and E. A. MacDougall-Shackleton. 2019. "Songbirds show odour-based discrimination of similarity and diversity at the major histocompatibility complex." *Animal Behaviour* 158:131–138.

Leclaire, S., M. Strandh, J. Mardon, H. Westerdahl, and F. Bonadonna. 2017. "Odour-based discrimination of similarity at the major histocompatibility complex in birds." *Proceedings of the Royal Society B: Biological Sciences* 284:20162466.

Gonzalez, G., G. Sorci, L. C. Smith, and F. de Lope. 2001. "Testosterone and sexual signalling in male house sparrows (*Passer domesticus*)." *Behavioral Ecology and Sociobiology* 50:557–562.

McGlothlin, J. W., J. Jawor, T. J. Greives, J. M. Casto, J. L. Phillips, and E. D. Ketterson. 2008. "Hormones and honest signals: Males with larger ornaments elevate testosterone more when challenged." *Journal of Evolutionary Biology* 21 (1): 39–48.

Gerald, M. S., and M. T. McGuire. 2007. "Secondary sexual coloration and CSF 5-HIAA are correlated in vervet monkeys (*Cercopithecus aethiops sabaeus*)." *Journal of Medical Primatology* 36 (6):348–354.

Amo, L., I. López-Rull, I. Pagán, and C. M. Garcia. 2012. "Male quality and conspecific scent preferences in the house finch, *Carpodacus mexicanus*." *Animal Behavior* 84 (6):1483–1489.

Rosvall, K. A. 2008. "Sexual selection on aggressiveness in females: Evidence from an experimental test with tree swallows." *Animal Behaviour* 75:1603–1610.

Rosvall, K. A., C. M. Bergeon Burns, J. Barske, J. L. Goodson, B. A. Schlinger, D. R. Sengelaub, and E. D. Ketterson. 2012. "Neural sensitivity to sex steroids predicts individual differences in aggression: Implications for behavioural evolution." *Proceedings of the Royal Society B: Biological Sciences* 279:3547–3555.

Whittaker, D. J., K. A. Rosvall, S. P. Slowinski, H. A. Soini, M. V. Novotny, and E. D. Ketterson. 2018. "Songbird chemical signals reflect uropygial gland androgen sensitivity and predict aggression: Implications for the role of the periphery in chemosignaling." *Journal of Comparative Physiology A: Sensory, Neural, and Behavioral Physiology* 204 (1):5–15.

Yeh, P. J. 2004. "Rapid evolution of a sexually selected trait following population establishment in a novel habitat." *Evolution* 58 (1):166–174.

Yeh, P. J., and T. D. Price. 2004. "Adaptive phenotypic plasticity and the successful colonization of a novel environment." *American Naturalist* 164 (4):531–542.

Atwell, J. W., G. C. Cardoso, D. J. Whittaker, T. D. Price, and E. D. Ketterson. 2014. "Hormonal, behavioral and life-history traits exhibit correlated shifts in relation to population establishment in a novel environment." *American Naturalist* 184 (6):E147–E160.

Newman, M. M., P. J. Yeh, and T. D. Price. 2006. "Reduced territorial responses in dark-eyed juncos following population establishment in a climatically mild environment." *Animal Behaviour* 71:893–899.

Atwell, J. W., G. C. Cardoso, D. J. Whittaker, K. W. Robertson, S. J. Campbell-Nelson, and E. D. Ketterson. 2012. "Boldness and stress physiology in a novel urban environment suggest rapid evolutionary adaptation." *Behavioral Ecology* 23:960–969.

Bonadonna, F., and G. A. Nevitt. 2004. "Partner-specific odor recognition in an Antarctic seabird." *Science* 306:835.

Whittaker, D. J., K. M. Richmond, A. K. Miller, R. Kiley, C. Bergeon Burns, J. W. Atwell, and E. D. Ketterson. 2011. "Intraspecific preen oil odor preferences in dark-eyed juncos (*Junco hyemalis*)." *Behavioral Ecology* 22:1256–1263.

CHAPTER 4. WHAT DOES SEXY SMELL LIKE?

McGlone, J. J., E. O. Aviles-Rosa, C. Archer, M. M. Wilson, K. D. Jones, E. M. Matthews, A. A. Gonzalez, and E. Reyes. 2020. "Understanding sow sexual behavior and the application of the boar pheromone to stimulate sow reproduction." In *Animal Reproduction in Veterinary Medicine*, edited by F. Aral, R. Payan-Carreira, and M. Quaresma. London: IntechOpen. https://doi.org/10.5772/intechopen.90774.

Keller, M., Q. Douhard, M. J. Baum, and J. Bakker. 2006. "Destruction of the main olfactory epithelium reduces female sexual behavior and olfactory investigation in female mice." *Chemical Senses* 31:315–323.

Macrides, F., A. Bartke, and S. Dalterio. 1975. "Strange females increase plasma testosterone levels in male mice." *Science* 189:1104–1106.

Bonilla-Jaime, H., G. Vázquez-Palacios, M. Arteaga-Silva, and S. Retana-Márquez. 2006. "Hormonal responses to different sexually related conditions in male rats." *Hormones and Behavior* 49:376–382.

Ziegler, T. E., N. J. Schultz-Darken, J. J. Scott, C. T. Snowdon, and C. F. Ferris. 2005. "Neuroendocrine response to female ovulatory odors depends upon social condition in male common marmosets, *Callithrix jacchus*." *Hormones and Behavior* 47:56–64.

Koyama, S., and S. Kamimura. 2000. "Influence of social dominance and female odor on the sperm activity of male mice." *Physiology & Behavior* 71:415–422.

Balthazart, J., and E. Schoffeniels. 1979. "Pheromones are involved in the control of sexual behaviour in birds." *Naturwissenschaften* 66:55–56.

Hirao, A., M. Aoyama, and S. Sugita. 2009. "The role of uropygial gland on sexual behavior in domestic chicken *Gallus gallus domesticus*." *Behavioural Processes* 80:115–120.

Balthazart, J., and M. Taziaux. 2009. "The underestimated role of olfaction in avian reproduction?" *Behavioral Brain Research* 200:248–259.

Douglas, H. D., III. 2008. "Prenuptial perfume: Alloanointing in the social rituals of the crested auklet (*Aethia cristatella*) and the transfer of arthropod deterrents." *Naturwissenschaften* 95:45–53.

Hagelin, J. C., I. L. Jones, and L. E. L. Rasmussen. 2003. "A tangerine-scented social odour in a monogamous seabird." *Proceedings of the Royal Society Biological Sciences Series B* 270:1323–1329.

Clark, G. A. 1970. "Avian bill-wiping." *The Wilson Bulletin* 82 (3):279–288.

Tinbergen, N. 1952. "'Derived' activities: Their causation, biological significance, origin, and emancipation during evolution." *Quarterly Review of Biology* 27 (1):1–32.

Maestripieri, D., G. Schino, F. Aureli, and A. Trioisi. 1992. "A modest proposal: Displacement activities as an indicator of emotions in primates." *Animal Behaviour* 44:967–979.

Thiessen, D. D., and A. E. Harriman. 1986. "Harderian gland exudates in the male *Meriones unguiculatus* regulate female proceptive behavior, aggression, and investigation." *Journal of Comparative Psychology* 100:85–87.

Wolff, J. O., M. H. Watson, and S. A. Thomas. 2002. "Is self-grooming by male prairie voles a predictor of mate choice?" *Ethology* 108:169–179.

Ferkin, Michael H., Stuart T. Leonard, Lori A. Heath, and Guillermo Paz-y-Miño C. 2001. "Self-grooming as a tactic used by prairie voles *Microtus ochrograster* to enhance sexual communication." *Ethology* 107:939–949.

Wiley, R. H. 1991. "Lekking in birds and mammals: Behavioral and evolutionary issues." *Advances in the Study of Behavior* 20:201–291.

DuVal, E. H. 2007. "Cooperative display and lekking behavior of the lance-tailed manakin (*Chiroxiphia lanceolata*)." *The Auk* 124 (4):1168–1185.

DuVal, E. H. 2007. "Adaptive advantages of cooperative courtship for subordinate male lance-tailed manakins." *The American Naturalist* 169 (4):423–432.

DuVal, E. H., and B. Kempenaers. 2008. "Sexual selection in a lekking bird: The relative opportunity for selection by female choice and male competition." *Proceedings of the Royal Society B* 275:1995–2003.

DuVal, E. H. 2012. "Variation in annual and lifetime reproductive success of lance-tailed manakins: Alpha experience mitigates effects of senescence on siring success." *Proceedings of the Royal Society B Biological Sciences* 279:1551–1559.

Sardell, R. J., and E. H. DuVal. 2014. "Differential allocation in a lekking bird: Females lay larger eggs and are more likely to have male chicks when they mate with less related males." *Proceedings of the Royal Society B* 281 (1774):20132386.

Marshall, R. C., K. L. Buchanan, and C. K. Catchpole. 2003. "Sexual selection and individual genetic diversity in a songbird." *Proceedings of the Royal Society B* 270:S248.

Oh, K. P., and A. V. Badyaev. 2006. "Adaptive genetic complementarity in mate choice coexists with selection for elaborate sexual traits." *Proceedings of the Royal Society of London B Biological Sciences* 273 (1596):1913–1919.

Bonadonna, F., and A. Sanz-Aguilar. 2012. "Kin recognition and inbreeding avoidance in wild birds: The first evidence for individual kin-related odour recognition." *Animal Behaviour* 84:509–513.

Coffin, H. R., J. V. Watters, and J. M. Mateo. 2011. "Odor-based recognition of familiar and related conspecifics: A first test conducted on captive Humboldt penguins (*Spheniscus humboldti*)." *PLoS ONE* 6 (9):e25002.

Krause, E. T., O. Krüger, P. Kohlmeier, and B. A. Caspers. 2012. "Olfactory kin recognition in a songbird." *Biology Letters* 8 (3):327–329.

Caspers, B. A., J. C. Hagelin, M. Paul, S. Bock, S. Willeke, and E. T. Krause. 2017. "Zebra finch chicks recognise parental scent, and retain chemosensory knowledge of their genetic mother, even after egg cross-fostering." *Scientific Reports* 7:12859.

Caspers, B., A. Gagliardo, and E. T. Krause. 2015. "Impact of kin odour on reproduction in zebra finches." *Behavioral Ecology and Sociobiology* 69:1827–1833.

Dunn, P. O., J. L. Bollmer, C. R. Freeman-Gallant, and L. A. Whittingham. 2013. "MHC variation is related to a sexually selected ornament, survival, and parasite resistance in common yellowthroats." *Evolution* 67 (3):679–687.

Slade, J. W. G., M. J. Watson, and E. A. MacDougall-Shackleton. 2017. "Birdsong signals individual diversity at the major histocompatibility complex." *Biology Letters* 13:20170430.

von Schantz, T., H. Wittzell, G. Goransson, M. Grahn, and K. Persson. 1996. "MHC genotype and male ornamentation: Genetic evidence for the Hamilton-Zuk model." *Proceedings of the Royal Society of London B Biological Sciences* 263: 265–271.

Strandh, M., H. Westerdahl, M. Pontarp, B. Canbäck, M.-P. Dubois, C. Miquel, P. Taberlet, and F. Bonadonna. 2012. "Major histocompatibility complex class II compatibility, but not class I, predicts mate choice in a bird with highly developed olfaction." *Proceedings of the Royal Society B* 279 (1746):4457–4463.

Gillingham, M. A. F., D. S. Richardson, H. Løvlie, A. Moynihan, K. Worley, and T. Pizzari. 2009. "Cryptic preference for MHC-dissimilar females in male red junglefowl, *Gallus gallus*." *Proceedings of the Royal Society of London B Biological Sciences* 276:1083–1092.

Yamaguchi, M., K. Yamazaki, G. K. Beauchamp, J. Bard, L. Thomas, and E. A. Boyse. 1981. "Distinctive urinary odors governed by the major histocompatibility locus of the mouse." *Proceedings of the National Academy of Sciences* 78 (9):5817–5820.

Milinski, M., S. Griffiths, K. M. Wegner, T. B. H. Reusch, A. Haas-Assenbaum, and T. Boehm. 2005. "Mate choice decisions of stickleback females predictably modified by MHC peptide ligands." *Proceedings of the National Academy of Sciences* 102:4414–4418.

Slade, J. W. G., M. J. Watson, T. R. Kelly, G. B. Gloor, M. A. Bernards, and E. A. MacDougall-Shackleton. 2016. "Chemical composition of preen wax reflects major histocompatibility complex similarity in songbirds." *Proceedings of the Royal Society B* 283:20161966. https://doi.org/10.1098/rspb.2016.1966.

Leclaire, S., W. F. D. van Dongen, S. Voccia, T. Merkling, C. Ducamp, S. A. Hatch, P. Blanchard, E. Danchin, and R. H. Wagner. 2014. "Preen secretions encode information on MHC similarity in certain sex-dyads in a monogamous seabird." *Scientific Reports* 4:6920. https://doi.org/10.1038/srep06920.

Grieves, L. A., G. B. Gloor, M. A. Bernards, and E. A. Macdougall-Shackleton. 2019. "Songbirds show odour-based discrimination of similarity and diversity at the major histocompatibility complex." *Animal Behaviour* 158:131–138.

Leclaire, S., M. Strandh, J. Mardon, H. Westerdahl, and F. Bonadonna. 2017. "Odour-based discrimination of similiarity at the major histocompatibility complex in birds." *Proceedings of the Royal Society B* 284:20162466.

Wyatt, T. D. 2009. "Fifty years of pheromones." *Nature* 457 (262–263).

Darwin, C. 1871. *The Descent of Man, and Selection in Relation to Sex.* London: J. Murray.

Whittaker, D. J., N. M. Gerlach, H. A. Soini, M. V. Novotny, and E. D. Ketterson. 2013. "Bird odour predicts reproductive success." *Animal Behaviour* 86:697–703.

CHAPTER 5. MAKING SCENTS OF BACTERIA

Gorman, M. L., D. B. Nedwell, and R. M. Smith. 1974. "An analysis of the contents of the anal and scent pockets of *Herpestes auropunctatus* (Carnivora: Viverridae)." *Journal of Zoology* 172 (3):389–399.

Albone, E. S., P. E. Gosden, G. C. Ware, D. W. MacDonald, and N. G. Hough. 1978. "Bacterial action and chemical signalling in the red fox (*Vulpes vulpes*) and other mammals." In *Flavor Chemistry of Animal Foods*, edited by R. W. Bullard, 78–91. Washington, DC: American Chemical Society.

Albone, E. S., P. E. Gosden, and G. C. Ware. 1977. "Bacteria as a source of chemical signals in mammals." In *Chemical Signals in Vertebrates*, edited by D. Müller-Schwarze and M. M. Mozell, 35–43. Boston: Springer.

Goodwin, T. E., L. J. Broederdorf, B. A. Burkert, I. H. Hirwa, D. B. Mark, Z. J. Waldrip, R. A. Kopper, M. V., et al. 2012. "Chemical signals of elephant musth: Temporal aspects of microbially-mediated modifications." *Journal of Chemical Ecology* 38:81–87.

Zechman, J. M., I. G. Martin, J. L. Wellington, and G. K. Beauchamp. 1984. "Perineal scent gland of wild and domestic cavies: Bacterial activity and urine as sources of biologically significant odors." *Physiology & Behavior* 32:269–274.

Voigt, C., B. Caspers, and S. Speck. 2005. "Bats, bacteria, and bat smell: Sex-specific diversity of microbes in a sexually selected scent organ." *Journal of Mammalogy* 86 (4):745–749.

Rennie, P. J., K. T. Holland, A. I. Mallet, W. J. Watkins, and D. B. Gower. 1989. "Testosterone-metabolism by human axillary bacteria." *Biochemical Society Transactions* 17:1017–1018.

Gower, D. B., K. T. Holland, A. I. Mallet, P. J. Rennie, and W. J. Watkins. 1994. "Comparison of 16-Androstene steroid concentrations in sterile apocrine sweat and axillary secretions: Interconversions of 16-Androstenes by the axillary microflora—a mechanism for axillary odour production in man?" *Journal of Steroid Biochemistry and Molecular Biology* 48 (4):409–418.

Shawkey, M. D., S. R. Pillai, and G. E. Hill. 2009. "Do feather-degrading bacteria affect sexually selected plumage color?" *Naturwissenschaften* 96:123–128.

Shawkey, M. D., S. R. Pillai, and G. E. Hill. 2003. "Chemical warfare? Effects of uropygial oil on feather-degrading bacteria." *Journal of Avian Biology* 34: 345–349.

Martín-Vivaldi, M., A. Peña, J. M. Peralta-Sánchez, L. Sánchez, S. Ananou, M. Ruiz-Rodríguez, and J. J. Soler. 2010. "Antimicrobial chemicals in hoopoe preen secretions are produced by symbiotic bacteria." *Proceedings of the Royal Society B: Biological Sciences* 277:123–130.

Soler, J. J., M. Martín-Vivaldi, J. M. Peralta-Sánchez, L. Arco, and N. Juárez-García-Pelayo. 2014. "Hoopoes color their eggs with antimicrobial uropygial secretions." *Naturwissenschaften* 101:697–705.

Lemfack, M. C., B.-O. Gohlke, S. M. T. Toguem, S. Preissner, B. Piechulla, and R. Preissner. 2017. "mVOC 2.0: A database of microbial volatiles." *Nucleic Acids Research* 46 (D1):D1261–D1265.

Whittaker, D. J., S. P. Slowinski, J. M. Greenberg, O. Alian, A. D. Winters, M. M. Ahmad, M. J. E. Burrell, H. A. Soini, M. V. Novotny, E. D. Ketterson, K. R. Theis. 2019. "Experimental evidence that symbiotic bacteria produce chemical cues in a songbird." *Journal of Experimental Biology* 222:jeb202978.

Zilber-Rosenberg, I., and E. Rosenberg. 2008. "Role of microorganisms in the evolution of animals and plants: The hologenome theory of evolution." *FEMS Microbiology Review* 32:723–735.

Theis, K. R., R. Romero, J. M. Greenberg, A. D. Winters, V. Garcia-Flores, K. Motomura, M. M. Ahmad, J. Galaz, M. Arenas-Hernandez, and N. Gomez-Lopez. 2020. "No consistent evidence for microbiota in murine placental and fetal tissues." *mSphere* 5:e00933-19.

Theis, K. R., R. Romero, A. D. Winters, A. H. Jobe, and N. Gomez-Lopez. 2020. "Lack of evidence for microbiota in the placental and fetal tissues of rhesus macaques." *mSphere* 5:e00210-20.

Theis, K. R., R. Romero, A. D. Winters, J. M. Greenberg, N. Gomez-Lopez, A. Alhousseini, J. Bieda, et al. 2019. "Does the human placenta delivered at term have a microbiota? Results of cultivation, quantitative real-time PCR, 16SrRNA gene sequencing, and metagenomics." *American Journal of Obstetrics & Gynecology* 220 (3):267.E1-267.E39.

Kirk, R. G. W. 2012. " 'Life in a germ-free world': Isolating life from the laboratory animal to the bubble boy." *Bulletin of the History of Medicine* 86:237–275.

Luczynski, P., K.-A. McVey Neufeld, C. S. Oriach, G. Clarke, T. G. Dinan, and J. F. Cryan. 2016. "Growing up in a bubble: Using germ-free animals to assess the influence of the gut microbiota on brain and behavior." *International Journal of Neuropsychopharmacology* 19 (8):1–17.

Sudo, N., Y. Chida, Y. Aiba, J. Sonoda, N. Oyama, X.-N. Yu, C. Kubo, and Y.Koga. 2004. "Postnatal microbial colonization programs the hypothalamic-pituitary-adrenal system for stress response in mice." *Journal of Physiology* 558:263–275.

CHAPTER 6. THANKS FOR SHARING

Bishai, D., L. Liu, S. Shiau, H. Wang, C. Tsai, M. Liao, S. Prakash, and T. Howard. 2011. "Quantifying school officials' exposure to bacterial pathogens at graduation ceremonies using repeated observational measures." *Journal of School Nursing* 27 (3):219–224.

Frumin, I., O. Perl, Y. Endevelt-Shapira, A. Eisen, N. Eshel, I. Heller, M. Shemesh, et al. 2015. "A social chemosignaling function for human handshaking." *eLife* 4:e05154.

Lax, S., D. P. Smith, J. Hampton-Marcell, S. M. Owens, K. M. Handley, N. M. Scott, S. M. Gibbons, et al. 2014. "Longitudinal analysis of microbial interaction between humans and the indoor environment." *Science* 345:1048–1052.

REFERENCES

Song, S. J., C. L. Lauber, E. K. Costello, C. A. Lozupone, G. Humphrey, D. Berg-Lyons, J. G. Caporaso, et al. 2013. "Cohabiting family members share microbiota with one another and with their dogs." *eLife* 2:e00458. https://doi.org/10.7554/eLife.00458.

Meadow, J. F., A. C. Bateman, K. M. Herkert, T. K. O'Connor, and J. L. Green. 2013. "Significant changes in the skin microbiome mediated by the sport of roller derby." *PeerJ* 1:e53. https://doi.org/10.7717/peerj.53.

Whittaker, D. J., N. M. Gerlach, S. P. Slowinski, K. P. Corcoran, A. D. Winters, H. A. Soini, M. V. Novotny, E. D. Ketterson, and K. R. Theis. 2016. "Social environment has a primary influence on the microbial and odor profiles of a chemically signaling songbird." *Frontiers in Ecology and Evolution* 4:90. https://doi.org/10.3389/fevo.2016.00090.

Ruiz-Rodríguez, M., J. J. Soler, M. Martín-Vivaldi, A. M. Martín-Platero, M. Méndez, J. M. Peralta-Sánchez, S. Ananou, E. Valdivia, and M. Martínez-Bueno. 2014. "Environmental factors shape the community of symbionts in the hoopoe uropygial gland more than genetic factors." *Applied and Environmental Microbiology* 80 (21):6714–6723.

Pearce, D. S., B. A. Hoover, S. Jennings, G. A. Nevitt, and K. M. Docherty. 2017. "Morphological and genetic factors shape the microbiome of a seabird species (*Oceanodroma leucorhoa*) more than environmental and social factors." *Microbiome* 5:146.

Kulkarni, S., and P. Heeb. 2007. "Social and sexual behaviours aid transmission of bacteria in birds." *Behavioural Processes* 74:88–92.

Tung, J., L. B. Barreiro, M. B. Burns, J.-C. Grenier, J. Lynch, L. E. Grieneisen, J. Altmann, et al. 2015. "Social networks predict gut microbiome composition in wild baboons." *eLife* 4:05224.

Moeller, A. H., S. Foerster, M. L. Wilson, A. E. Pusey, B. H. Hahn, and H. Ochman. 2016. "Social behavior shapes the chimpanzee pan-microbiome." *Science Advances* 2:e1500997.

Lombardo, M. P. 2008. "Access to mutualistic endosymbiotic microbes: An underappreciated benefit of group living." *Behavioral Ecology and Sociobiology* 62:479–497.

Zeh, J. A., and D. W. Zeh. 1996. "The evolution of polyandry 1: Intragenomic conflict and genetic incompatibility." *Proceedings of the Royal Society B: Biological Sciences* 263:1711–1717.

Gerlach, N. M., J. W. McGlothlin, P. G. Parker, and E. D. Ketterson. 2012. "Promiscuous mating produces offspring with higher lifetime fitness." *Proceedings of the Royal Society B: Biological Sciences* 279:860–866.

Lombardo, M. P., P. A. Thorpe, and H. W. Power. 1999. "The beneficial sexually transmitted microbe hypothesis of avian copulation." *Behavioral Ecology* 10 (3):333–350.

Dunbar, R. I. M. 1998. "The social brain hypothesis." *Evolutionary Anthropology* 6:178–190.

Troyer, K. 1984. "Microbes, herbivory, and the evolution of social behavior." *Journal of Theoretical Biology* 106:157–169.

Harris, R. N., and D. E. Gill. 1980. "Communal nesting, brooding behavior, and embryonic survival of the four-toed salamander *Hemidactylium scutatum*." *Herpetologica* 36 (2):141–144.

Banning, J. L., A. L. Weddle, G. W. Wahl III, M. A. Simon, A. Lauer, R. L. Walters, and R. Harris. 2008. "Antifungal skin bacteria, embryonic survival, and communal nesting in four-toed salamanders, Hemidactylium scutatum." Oecologia 156:423–429.

Koch, H., and P. Schmid-Hempel. 2011. "Socially transmitted gut microbiota protect bumble bees against an intestinal parasite." Proceedings of the National Academy of Sciences 108 (48):19288–19292.

Brown, M. J. F., R. Schmid-Hempel, and P. Schmid-Hempel. 2003. "Strong context-dependent virulence in a host-parasite system: Reconciling genetic evidence with theory." Journal of Animal Ecology 72 (6):994–1002.

Montiel-Castro, A. J., R. M. González-Cervantes, G. Bravo-Ruiseco, and G. Pacheco-López. 2013. "The microbiota-gut-brain axis: Neurobehavioral correlates, health and sociality." Frontiers in Integrative Neuroscience 7:70.

Lewin-Epstein, O., R. Aharonov, and L. Hadany. 2017. "Microbes can help explain the evolution of altruism." Nature Communications 8:14040.

Burns, A. R., E. Miller, M. Agarwal, A. Rolig, K. Milligan-Myhre, S. Seredick, K. Guillemin, and B. J. M. Bohannon. 2017. "Interhost dispersal alters microbiome assembly and can overwhelm host innate immunity in an experimental zebrafish model." Proceedings of the National Academy of Sciences 114 (42):11181–11186.

Love, S. 2020. "Your body odor might change in quarantine." Vice, April 30, 2020. https://www.vice.com/en/article/xgqeva/your-body-odor-might-change-in-coronavirus-quarantine.

CHAPTER 7. MHC: MAGICAL HAPPINESS CONTROLLER?

Shiina, T., K. Hosomichi, H. Inoko, and J. K. Kulski. 2009. "The HLA genomic loci map: Expression, interaction, diversity and disease." Journal of Human Genetics 54:15–39.

Yamazaki, K., E. A. Boyse, V. Miké, H. T. Thaler, B. J. Mathieson, J. Abbott, J. Boyse, Z. A. Zayas, and L. Thomas. 1976. "Control of mating preferences in mice by genes in the major histocompatibility complex." Journal of Experimental Medicine 144:1324–1335.

Wedekind, C., T. Seebeck, F. Bettens, and A. J. Paepke. 1995. "MHC-dependent mate preferences in humans." Proceedings of the Royal Society of London B: Biological Sciences 260:245–249.

Aldhous, P. 1995. "Darling, you smell wonderfully different." New Scientist, issue 1976, May 6, 1995.

Chaix, R., C. Cao, and P. Donnelly. 2008. "Is mate choice in humans MHC-dependent?" PLOS Genetics 4 (9):e1000184.

Wyatt, T. D. 2020. "Reproducible research into human chemical communication by cues and pheromones: Learning from psychology's renaissance." Philosophical Transactions of the Royal Society B: Biological Sciences 375:20190262.

Wyatt, T. 2013. "The smelly mystery of the human pheromone." TED, September 2013. https://www.ted.com/talks/tristram_wyatt_the_smelly_mystery_of_the_human_pheromone.

Martin, A., M. Saathoff, F. Kuhn, H. Max, L. Terstegen, and A. Natsch. 2010. "A functional ABCC11 allele is essential in the biochemical formation of human axillary odor." Journal of Investigative Dermatology 130:529–540.

Mansky, J. 2018. "The dubious science of genetics-based dating." *Smithsonian Magazine* (website), February 14, 2018. https://www.smithsonianmag.com/science -nature/dubious-science-genetics-based-dating-180968151/.

Kaufman, J., S. Milne, T. W. F. Gobel, B. A. Walker, J. P. Jacob, C. Auffray, R. Zoorob, and S. Beck. 1999. "The chicken B locus is a minimal essential major histocompatibility complex." *Nature* 401:923–925.

Bonneaud, C., G. Sorci, V. Morin, H. Westerdahl, R. Zoorob, and H. Wittzell. 2004. "Diversity of Mhc class I and IIB genes in house sparrows (*Passer domesticus*)." *Immunogenetics* 55:855–865.

Westerdahl, H., H. Wittzell, and T. von Schantz. 2000. "Mhc diversity in two passerine birds: No evidence for a minimal essential Mhc." *Immunogenetics* 52:92–100.

Edwards, S. V., M. Grahn, and W. K. Potts. 1995. "Dynamics of Mhc evolution in birds and crocodilians: Amplification of class II genes with degenerate primers." *Molecular Ecology* 4:719–729.

Edwards, S. V., J. Gasper, D. Garrigan, D. Martindale, and B. F. Koop. 2000. "A 39-kb sequence around a blackbird Mhc class II gene: Ghost of selection past and songbird genome architecture." *Molecular Biology and Evolution* 17 (9):1384–1395.

Edwards, S. V., J. Gasper, and M. March. 1998. "Genomics and polymorphism of Agph-DAB1, an Mhc class II b gene in red-winged blackbirds (*Agelaius phoeniceus*)." *Molecular Biology and Evolution* 15 (3):236–250.

Balakrishnan, C. N., R. Ekblom, M. Volker, H. Westerdahl, R. Godinez, H. Kotkiewicz, D. W. Burt, et al. 2010. "Gene duplication and fragmentation in the zebra finch major histocompatibility complex." *BMC Biology* 8:29.

Zagalska-Neubauer, M., W. Babik, M. Stuglik, L. Gustafsson, M. Cichoń, and J. Radwan. 2010. "454 sequencing reveals extreme complexity of the class II major histocompatibility complex in the collared flycatcher." *BMC Evolutionary Biology* 10:395.

Bollmer, J. L., P. O. Dunn, L. A. Whittingham, and C. Wimpee. 2010. "Extensive MHC class II B gene duplication in a passerine, the common yellowthroat (*Geothlypis trichas*)." *Journal of Heredity* 101 (4):448–460.

Drews, A., and H. Westerdahl. 2019. "Not all birds have a single dominantly expressed MHC-I gene: Transcription suggests that siskins have many highly expressed MHC-I genes." *Scientific Reports* 9:19506.

Leinders-Zufall, T., P. A. Brennan, P. Widmayer, P. Chandramani, A. Maul-Pavicic, M. Jager, X.-H. Li, et al. 2004. "MHC class I peptides as chemosensory signals in the vomeronasal organ." *Science* 306:1033–1037.

Milinski, M., S. Griffiths, K. M. Wegner, T. B. H. Reusch, A. Haas-Assenbaum, and T. Boehm. 2005. "Mate choice decisions of stickleback females predictably modified by MHC peptide ligands." *Proceedings of the National Academy of Sciences* 102:4414–4418.

Meredith, M. 2001. "Human vomeronasal organ function: A critical review of best and worst cases." *Chemical Senses* 26 (4):433–445.

Halpern, M., and A. Martínez-Marcos. 2003. "Structure and function of the vomeronasal system: An update." *Progress in Neurobiology* 70 (3):245–318.

Bolnick, D. I., L. K. Snowberg, J. G. Caporaso, C. L. Lauber, R. Knight, and W. E. Stutz. 2014. "Major histocompatibility complex class IIb polymorphism influences gut microbiota composition and diversity." *Molecular Ecology* 23:4831–4845.

De Palma, G., A. Capilla, I. Nadal, E. Nova, T. Pozo, V. Varea, I. Polanco, et al. 2010. "Interplay between human leukocyte antigen genes and the microbial colonization process of the newborn intestine." *Current Issues in Molecular Biology* 12 (1):1–10.

Kubinak, J. L., W. X. Stephens, R. Soto, C. Petersen, T. Chiaro, L. Gogokhia, R. Bell, et al. 2015. "MHC variation sculpts individualized microbial communities that control susceptibility to enteric infection." *Nature Communications* 6:8642.

Williams, M. R., and R. L. Gallo. 2015. "The role of the skin microbiome in atopic dermatitis." *Current Allergy and Asthma Reports* 15:65.

Hernández-Gómez, O., J. T. Briggler, and R. N. Williams. 2018. "Influence of immunogenetics, sex and body condition on the cutaneous microbial communities of two giant salamanders." *Molecular Ecology* 27 (8):1915–1929.

Leclaire, S., M. Strandh, G. Dell'Ariccia, M. Gabirot, H. Westerdahl, and F. Bonadonna. 2019. "Plumage microbiota covaries with major histocompatibility complex in blue petrels." *Molecular Ecology* 28:833–846.

Bushdid, C., M. O. Magnasco, L. B. Vosshall, and A. Keller. 2014. "Humans can discriminate more than 1 trillion olfactory stimuli." *Science* 343 (6177):1370–1372.

Younger, R. M., C. Amadou, G. Bethel, A. Ehlers, K. F. Lindahl, S. Forbes, R. Horton, et al. 2001. "Characterization of clustered MHC-linked olfactory receptor genes in human and mouse." *Genome Research* 11 (4):519–530.

Miller, M. M., C. M. Robinson, J. Abernathy, R. M. Goto, M. K. Hamilton, H. Zhou, and M. E. Delaney. 2013. "Mapping genes to chicken microchromosome 16 and discovery of olfactory and scavenger receptor genes near the major histocompatibility complex." *Journal of Heredity* 105:203–215.

Milinski, M., and C. Wedekind. 2001. "Evidence for MHC-correlated perfume preferences in humans." *Behavioral Ecology* 12 (2):140–149.

Bonneaud, C., O. Chastel, P. Federici, H. Westerdahl, and G. Sorci. 2006. "Complex Mhc-based mate choice in a wild passerine." *Proceedings of the Royal Society B: Biological Sciences* 273:1111–1116.

Milinski, M., S. Griffiths, K. M. Wegner, T. B. H. Reusch, A. Haas-Assenbaum, and T. Boehm. 2005. "Mate choice decisions of stickleback females predictably modified by MHC peptide ligands." *Proceedings of the National Academy of Sciences* 102:4414–4418.

Aeschlimann, P. B., M. A. Haberli, T. B. H. Reusch, T. Boehm, and M. Milinski. 2003. "Female sticklebacks Gasterosteus aculeatus use self-reference to optimize MHC allele number during mate selection." *Behavioral Ecology and Sociobiology* 54:119–126.

Fugger, L., and A. Svejgaard. 2000. "Association of MHC and rheumatoid arthritis: HLA-DR4 and rheumatoid arthritis: Studies in mice and men." *Arthritis Research* 2 (3):208–211.

International Multiple Sclerosis Genetics Consortium. 2015. "Class II HLA interactions modulate genetic risk for multiple sclerosis." *Nature Genetics* 47:1107–1113.

Nejentsev, S., J. M. M. Howson, N. M. Walker, J. Szeszko, S. F. Field, H. E. Stevens, P. Reynolds, et al. 2007. "Localization of type 1 diabetes susceptibility to the MHC class I genes HLA-B and HLA-A." *Nature* 450 (7171):887–892.

Knafler, G. J., J. A. Clark, P. D. Boersma, and J. L. Bouzat. 2012. "MHC diversity and mate choice in the Magellanic penguin, *Spheniscus magellanicus*." *Journal of Heredity* 103 (6):759–768.

Ekblom, R., A. Saether, M. Grahn, P. Fiske, A. Kalas, and J. Hoglund. 2004. "Major histocompatibility complex variation and mate choice in a lekking bird, the great snipe." *Molecular Ecology* 13:3821–3828.

Ober, C., L. R. Weitkamp, N. Cox, H. Dytch, D. Kostyu, and S. Elias. 1997. "HLA and mate choice in humans." *American Journal of Human Genetics* 61 (3):497–504.

Milinski, M. 2006. "The major histocompatibility complex, sexual selection, and mate choice." *Annual Review of Ecology and Systematics* 37:159–186.

Croy, I., G. Ritschel, D. Kreßner-Kiel, L. Schäfer, T. Hummel, J. Havlíček, J. Sauter, G. Ehninger, and A. H. Schmidt. 2020. "Marriage does not relate to major histocompatibility complex: A genetic analysis based on 3691 couples." *Proceedings of the Royal Society B: Biological Sciences* 287:20201800.

Rasner, C. A., P. Yeh, L. S. Eggert, K. E. Hunt, D. S. Woodruff, and T. D. Price. 2004. "Genetic and morphological evolution following a founder event in the dark-eyed junco, Junco hyemalis thurberi." *Molecular Ecology* 13:671–681.

Whittaker, D. J., A. L. Dapper, M. P. Peterson, J. W. Atwell, and E. D. Ketterson. 2012."Maintenance of MHC class IIB diversity in a recently established songbird population." *Journal of Avian Biology* 43 (2):109–118.

Richardson, D. S., J. Komdeur, T. Burke, and T. von Schantz. 2005. "MHC-based patterns of social and extra-pair mate choice in the Seychelles warbler." *Proceedings of the Royal Society B: Biological Sciences* 272:759–767.

Freeman-Gallant, C. R., M. Meguerdichian, N. T. Wheelwright, and S. V. Sollecito. 2003. "Social pairing and female mating fidelity predicted by restriction fragment length polymorphism similarity at the major histocompatibility complex in a songbird." *Molecular Ecology* 12:3077–3083.

CHAPTER 8. GIRL POWER

Amundsen, T. 2000. "Why are female birds ornamented?" *Trends in Ecology & Evolution* 15 (4):149–155.

West-Eberhard, M. J. 1983. "Sexual selection, social competition, and speciation." *Quarterly Review of Biology* 58 (2):155–183.

Burley, N. 1977. "Parental investment, mate choice, and mate quality." *Proceedings of the National Academy of Sciences* 74 (8):3476–3479.

Zhang, J.-X., W. Wei, J.-H. Zhang, and W.-H. Yang. 2010. "Uropygial gland-secreted alkanols contribute to olfactory sex signals in budgerigars." *Chemical Senses* 35 (5):375–382.

Mardon, J., S. M. Saunders, and F. Bonadonna. 2011. "Comments on recent work by Zhang and colleagues: 'Uropygial gland-secreted alkanols contribute to olfactory sex signals in budgerigars.'" *Chemical Senses* 36:3–4.

Whittaker, D. J, and J. C. Hagelin. 2021. "Female-based patterns and social function in avian chemical communication." *Journal of Chemical Ecology* 47 (1):43–62.

Jacob, S., A. Immer, S. Leclaire, N. Parthuisot, C. Ducamp, G. Espinasse, and P. Heeb. 2014. "Uropygial gland size and composition varies according to experimentally modified microbiome in great tits." *BMC Evolutionary Biology* 14:134.

Golüke, S., and B. Caspers. 2017. "Sex-specific differences in preen gland size of zebra finches during the course of breeding." *The Auk* 134 (4):821–831.

Martín-Vivaldi, M., M. Ruiz-Rodríguez, J. J. Soler, J. M. Peralta-Sánchez, M. Méndez, E. Valdivia, A. M. Martín-Platero, and M. Martínez-Bueno. 2009. "Seasonal, sexual and developmental differences in hoopoe Upupa epops preen gland morphology and secretions: Evidence for a role of bacteria." *Journal of Avian Biology* 40:191–205.

Møller, A. P., and K. Laursen. 2019. "Function of the uropygial gland in eiders (*Somateria mollissima*)." *Avian Research* 10:24.

Møller, A. P., G. Á. Czirják, and Philipp Heeb. 2009. "Feather microorganisms and uropygial antimicrobial defences in a colonial passerine bird." *Functional Ecology* 23 (6):1097–1102.

Coombes, H. A., P. Stockley, and J. L. Hurst. 2018. "Female chemical signalling underlying reproduction in mammals." *Journal of Chemical Ecology* 44:851–873.

Trivers, R. L. 1972. "Parental investment and sexual selection." In *Sexual Selection and the Descent of Man*, edited by B. Campbell, 52–97. Chicago: Aldine.

Hirao, A., M. Aoyama, and S. Sugita. 2009. "The role of uropygial gland on sexual behavior in domestic chicken *Gallus gallus domesticus*." *Behavioural Processes* 80:115–120.

Bonilla-Jaime, H., G. Vázquez-Palacios, M. Arteaga-Silva, and S. Retana-Márquez. 2006. "Hormonal responses to different sexually related conditions in male rats." *Hormones and Behavior* 49:376–382.

Ziegler, T. E., N. J. Schultz-Darken, J. J. Scott, C. T. Snowdon, and C. F. Ferris. 2005. "Neuroendocrine response to female ovulatory odors depends upon social condition in male common marmosets, *Callithrix jacchus*." *Hormones and Behavior* 47:56–64.

Cerda-Molina, A. L., L. Hernández-López, R. Chavira, M. Cárdenas, D. Paez-Ponce, H. Cervantes-De la Luz, and R. Mondragón-Ceballos. 2006. "Endocrine changes in male stumptailed macaques (*Macaca arctoides*) as a response to odor stimulation with vaginal secretions." *Hormones and Behavior* 49:81–87.

Miller, S. L., and J. K. Maner. 2009. "Scent of a woman: Men's testosterone responses to olfactory ovulation cues." *Psychological Science* 21 (2):276–283.

Whittaker, D. J., H. A. Soini, N. M. Gerlach, A. L. Posto, M. V. Novotny, and E. D. Ketterson. 2011. "Role of testosterone in stimulating seasonal changes in a potential avian chemosignal." *Journal of Chemical Ecology* 37:1349–1357.

delBarco-Trillo, J., C. R. Sacha, G. R. Dubay, and C. M. Drea. 2012. "*Eulemur, me lemur*: The evolution of scent-signal complexity in a primate clade." *Philosophical Transactions of the Royal Society B: Biological Sciences* 367:1909–1922.

Woszczylo, M., T. Jezierski, A. Szumny, W. Niżański, and M. Dzięcioł. 2020. "The role of urine in semiochemical communication between females and males of domestic dog (*Canis familiaris*) during estrus." *Animals* 10 (11):2112.

Drea, C. M., and E. S. Scordato. 2008. "Olfactory communication in the ringtailed lemur (*Lemur catta*): Form and function of multimodal signals." In *Chemical Signals in Vertebrates* 11, edited by J. L. Hurst, R. J. Beynon, S. C. Roberts, and T. D. Wyatt, 91–102. New York: Springer.

Huck, U. W., R. D. Lisk, S. Kim, and A. B. Evans. 1989. "Olfactory discrimination of estrous condition by the male golden hamster (*Mesocricetus auratus*)." *Behavioral and Neural Biology* 51:1–10.

Swaisgood, R. R., D. G. Lindburg, and H. Zhang. 2002. "Discrimination of oestrous status in giant pandas (*Ailuropoda melanoleuca*) via chemical cues in urine." *Journal of Zoology* 257:381–386.

Nie, Y., R. R. Swaisgood, Z. Zhang, Y. Hu, Y. Ma, and F. Wei. 2012. "Giant panda scent-marking strategies in the wild: Role of season, sex and marking surface." *Animal Behaviour* 84 (1):39–44.

Smith, J. L. D., C. McDougal, and D. Miquelle. 1989. "Scent marking in free-ranging tigers, *Panthera tigris*." *Animal Behaviour* 37:1–10.

Johnston, R. E. 1977. "The causation of two scent-marking behaviour patterns in female hamsters (*Mesocricetus auratus*)." *Animal Behaviour* 25:317–327.

Kuukasjärvi, S., C. Eriksson, E. Koskela, T. Mappes, K. Nissinen, and M. Rantala. 2004. "Attractiveness of women's body odors over the menstrual cycle: The role of oral contraceptives and receiver sex." *Behavioral Ecology* 15:579–584.

Boulet, M., J. C. Crawford, M. J. E. Charpentier, and C. M. Drea. 2010. "Honest olfactory ornamentation in a female-dominant primate." *Journal of Evolutionary Biology* 23:1558–1563.

Arakawa, H., S. Cruz, and T. Deak. 2011. "From models to mechanisms: Odorant communication as a key determinant of social behavior in rodents during illness-associated states." *Neuroscience and Biobehavioral Reviews* 35:1916–1928.

Kavaliers, M., E. Choleris, A. Agmo, and D. W. Pfaff. 2004. "Olfactory-mediated parasite recognition and avoidance: Linking genes to behavior." *Hormones and Behavior* 46:272–283.

Ferkin, M. H., E. S. Sorokin, R. E. Johnston, and C. J. Lee. 1997. "Attractiveness of scents varies with protein content of the diet in meadow voles." *Animal Behavior* 53:133–141.

Gillingham, M. A. F., D. S. Richardson, H. Løvlie, A. Moynihan, K. Worley, and T. Pizzari. 2009. "Cryptic preference for MHC-dissimilar females in male red junglefowl, *Gallus gallus*." *Proceedings of the Royal Society B: Biological Sciences* 276:1083–1092.

Leclaire, S., M. Strandh, J. Mardon, H. Westerdahl, and F. Bonadonna. 2017. "Odour-based discrimination of similarity at the major histocompatibility complex in birds." *Proceedings of the Royal Society B: Biological Sciences* 284:20162466.

West-Eberhard, M. J. 1983. "Sexual selection, social competition, and speciation." *Quarterly Review of Biology* 58 (2):155–183.

Trail, P. W. 1990. "Why should lek breeders be monomorphic?" *Evolution* 44:1837–1852.

Rosvall, K. A. 2008. "Sexual selection on aggressiveness in females: Evidence from an experimental test with tree swallows." *Animal Behaviour* 75:1603–1610.

Webb, W. H., D. H. Brunton, J. D. Aguirre, D. B. Thomas, M. Valcu, and J. Dale. 2016. "Female song occurs in songbirds with more elaborate female coloration and reduced sexual dimorphism." *Frontiers in Ecology and Evolution* 4:22. https://doi.org/10.3389/fevo.2016.00022.

DeVries, M. S., C. P. Winters, and J. M. Jawor. 2020. "Similarities in expression of territorial aggression in breeding pairs of northern cardinals, *Cardinalis cardinalis*." *Journal of Ethology* 38:377–382.

Arcese, P., P. K. Stoddard, and S. M. Hiebert. 1988. "The form and function of song in female song sparrows." *The Condor* 90 (1):44–50.

Cain, K. E., A. Cockburn, and N. E. Langmore. 2015. "Female song rates in response to simulated intruder are positively related to reproductive success." *Frontiers in Ecology and Evolution* 3:119.

Whittaker, D. J., K. A. Rosvall, S. P. Slowinski, H. A. Soini, M. V. Novotny, and E. D. Ketterson. 2018. "Songbird chemical signals reflect uropygial gland androgen sensitivity and predict aggression: Implications for the role of the periphery in chemosignaling." *Journal of Comparative Physiology A Sensory Neural and Behavioral Physiology* 204 (1):5–15.

Barrett, J., D. H. Abbott, and L. M. George. 1990. "Sensory cues and the suppression of reproduction in subordinate female marmoset monkeys, *Callithrix jacchus*." *Journal of Reproduction and Fertility* 97:301–310.

Abbott, D. H., W. Saltzman, N. J. Schultz-Darken, and T. E. Smith. 1997. "Specific neuroendocrine mechanisms not involving generalized stress mediate social regulation of female reproduction in cooperatively breeding marmoset monkeys." *Annals of the New York Academy of Sciences* 807:219–238.

Koenig, W. D., J. Haydock, and M. T. Stanback. 1998. "Reproductive roles in the cooperatively breeding acorn woodpecker: Incest avoidance versus reproductive competition." *American Naturalist* 151 (3):243–255.

Nelson-Flower, M. J., P. A. R. Hockey, C. O'Ryan, S. English, A. M. Thompson, K. Bradley, R. Rose, and A. R. Ridley. 2013. "Costly reproductive competition between females in a monogamous cooperatively breeding bird." *Proceedings of the Royal Society B: Biological Sciences* 280:20130728.

Reyer, H.-U., J. P. Dittami, and M. R. Hall. 1986. "Avian helpers at the nest: Are they psychologically castrated?" *Ethology* 71:216–228.

Rowley, I. 1978. "Communal activities among white-winged choughs *Corcorax melanorhamphus*." *Ibis* 120:178–197.

Brown, J. L. 1978. "Avian communal breeding systems." *Annual Review of Ecology and Systematics* 9:123–155.

Mays, N. A., C. M. Vleck, and J. Dawson. 1991. "Plasma luteinizing hormone, steroid hormones, behavioral role, and nest stage in the cooperatively breeding Harris' hawk (*Parabuteo unicinctus*)." *The Auk* 108:619–637.

Schoech, S. J., R. L. Mumme, and J. C. Wingfield. 1997. "Corticosterone, reproductive status, and body mass in a cooperative breeder, the Florida scrub-jay (*Aphelocoma coerulescens*)." *Physiological Zoology* 70 (1):68–73.

Wingfield, J. C., R. E. Hegner, and D. M. Lewis. 1991. "Circulating levels of luteinizing hormone and steroid hormones in relation to social status in the cooperatively breeding white-browed sparrow weaver, *Plocepasser mahali*." *Journal of Zoology* 225:43–58.

Kappeler, P. M. 1998. "To whom it may concern: The transmission and function of chemical signals in *Lemur catta*." *Behavioral Ecology and Sociobiology* 42:411–421.

Martín-Vivaldi, M., A. Peña, J. M. Peralta-Sánchez, L. Sánchez, S. Ananou, M. Ruiz-Rodríguez, and J. J. Soler. 2010. "Antimicrobial chemicals in hoopoe preen secretions are produced by symbiotic bacteria." *Proceedings of the Royal Society B: Biological Sciences* 277:123–130.

Martín-Vivaldi, M., J. J. Soler, J. M. Peralta-Sánchez, L. Arco, A. M. Martín-Platero, M. Martínez-Bueno, M. Ruiz-Rodríguez, and E. Valdivia. 2014. "Special structures of hoopoe eggshells enhance the adhesion of symbiont-carrying uropygial secretion that increase hatching success." *Journal of Animal Ecology* 83 (6):1289–1301.

Reneerkens, J., T. Piersma, and J. S. Sinninghe Damsté. 2002. "Sandpipers (Scolopacidae) switch from monoester to diester preen waxes during courtship and incubation, but why?" *Proceedings of the Royal Society B: Biological Sciences* 269:2135–2139.

Kolattukudy, P. E., S. Bohnet, and L. Rogers. 1987. "Diesters of 3-hydroxy fatty acids produced by the uropygial glands of female mallards uniquely during the mating season." *Journal of Lipid Research* 28:582–588.

Reneerkens, J., T. Piersma, and J. S. Sinninghe Damsté. 2005. "Switch to diester preen waxes may reduce avian nest predation by mammalian predators using olfactory cues." *Journal of Experimental Biology* 208:4199–4202.

Soini, H. A., S. E. Schrock, K. E. Bruce, D. Wiesler, E. D. Ketterson, and M. V. Novotny. 2007. "Seasonal variation in volatile compound profiles of preen gland secretions of the dark-eyed junco (*Junco hyemalis*)." *Journal of Chemical Ecology* 33 (1):183–198.

Cohen, J. 1981. "Olfaction and parental behavior in ring doves." *Biochemical Systematics and Ecology* 9 (4):351–354.

Golüke, S., S. Dörrenberg, E. T. Krause, and B. A. Caspers. 2016. "Female zebra finches smell their eggs." *PLOS One* 11 (5):e0155513.

Whittaker, D. J., N. M. Gerlach, S. P. Slowinski, K. P. Corcoran, A. D. Winters, H. A. Soini, M. V. Novotny, E. D. Ketterson, and K. R. Theis. 2016. "Social environment has a primary influence on the microbial and odor profiles of a chemically signaling songbird." *Frontiers in Ecology and Evolution* 4:90. https://doi.org/10.3389/fevo.2016.00090.

Kohlwey, S., E. T. Krause, M. C. Baier, C. Müller, and B. Caspers. 2016. "Chemical analyses reveal family-specific nest odor profiles in zebra finches (*Taeniopygia guttata*): A pilot study." In *Chemical Signals in Vertebrates* 13, edited by Bruce A. Schulte, T. E. Goodwin, and Michael H. Ferkin, 167–175. Cham, Switzerland: Springer.

Bonadonna, F., G. B. Cunningham, P. Jouventin, F. Hesters, and G. A. Nevitt. 2003. "Evidence for nest-odour recognition in two species of diving petrel." *Journal of Experimental Biology* 206:3719–3722.

Krause, E. T., and B. A. Caspers. 2012. "Are olfactory cues involved in nest recognition in two social species of estrildid finches?" *PLOS One* 7 (5):e36615.

Webster, B., W. Hayes, and T. W. Pike. 2015. "Avian egg odour encodes information on embryo sex, fertility and development." *PLOS One* 10 (1):e0116345.

Costanzo, A., S. Panseri, A. Giorgi, A. Romano, M. Caprioli, and N. Saino. 2016. "The odour of sex: Sex-related differences in volatile compound composition among barn swallow eggs carrying embryos of either sex." *PLOS One* 11 (11):e0165055. https://doi.org/10.1371/journal.pone.0165055.

Caspers, B. A., J. C. Hagelin, M. Paul, S. Bock, S. Willeke, and E. T. Krause. 2017. "Zebra finch chicks recognise parental scent, and retain chemosensory knowledge of their genetic mother, even after egg cross-fostering." *Scientific Reports* 7:12859.

Hagelin, J. C., J. C. Simonet, and T. R. Lyson. 2013. "Embryonic domestic chickens can detect compounds in an avian chemosignal before breathing air." In *Chemical Signals in Vertebrates* 12, edited by M. L. East and M. Dehnhard, 363–377. New York: Springer.

Caspers, B. A., and E. T, Krause. 2010. "Odour-based natal nest recognition in the zebra finch (*Taeniopygia guttata*), a colony-breeding songbird." *Biology Letters* 7:184–186.

Mínguez, E. 1997. "Olfactory nest recognition by British storm-petrel chicks." *Animal Behaviour* 53 (4):701–707.

Krause, E. T., O. Krüger, P. Kohlmeier, and B. A. Caspers. 2012. "Olfactory kin recognition in a songbird." *Biology Letters* 8 (3):327–329.

Caspers, B. A., J. I. Hoffman, P. Kohlmeier, O. Krüger, and E. T. Krause. 2013. "Olfactory imprinting as a mechanism for nest odour recognition in zebra finches." *Animal Behaviour* 86:85–90.

REFERENCES

Flicker, E. 2003. "Between brains and breasts—women scientists in fiction film: On the marginalization and sexualization of scientific competence." *Public Understanding of Science* 12:307–318.

INDEX

. . . .

This book is set in 10.25/15pt Literata, an old-style serif typeface designed by the type foundry TypeTogether. Literata was originally designed and released in 2015 for the Google Play Books application and later became available as open-source, with over 650 new characters for greater language support. The foundry writes:

> TypeTogether arrived at a hybrid solution that took inspiration from both Scotch and oldstyle Roman types. The resulting letterforms create a pleasant organic texture that helps to deliver very good results for ease of reading and comfort. The secondary style is an upright italic, meaning the lettershapes have an italicised construction and no slant to speak of.*

*Source: https://github.com/googlefonts/literata